22 IDEAS THAT SAVED THE ENGLISH COUNTRYSIDE

22 IDEAS THAT SAVED THE ENGLISH COUNTRYSIDE

THE CAMPAIGN TO PROTECT RURAL ENGLAND

PETER WAINE
OLIVER HILLIAM
FOREWORD BY ANDREW MOTION

F
FRANCES LINCOLN

To Juliette and Xanthe
Simply the best
P.W.

To Sara
My inspiration
O.H.

Contents

Foreword 6
Introduction 8

Idea 1	Discovering the Landscape	14
Idea 2	Preservation	24
Idea 3	The National Trust	34
Idea 4	Creating CPRE	42
Idea 5	Taming the Octopus	50
Idea 6	Keeping Villages Alive	58
Idea 7	Rural Planning	66
Idea 8	Democratic Planning	74
Idea 9	Green Belts	82
Idea 10	National Parks	94
Idea 11	The Sense Sublime	108
Idea 12	The Green and Pleasant Land	118
Idea 13	A Countryside Worth Fighting For	130
Idea 14	Urban Regeneration	140
Idea 15	Going Underground	148
Idea 16	Anti-Litter	154
Idea 17	The Right to Roam	162
Idea 18	Saving Our Forests	172
Idea 19	Nature Reserves	178
Idea 20	Cutting the Clutter	186
Idea 21	The Coast is the Countryside	198
Idea 22	The Country Code	206

Postscript: Unfinished Business 214
Index 218
Picture acknowledgements 223

Farndale in the North York Moors National Park.

Foreword
Andrew Motion

OPPOSITE
The iconic barns and dry stone walls of the Yorkshire Dales. Few landscapes give a clearer indication of the positive role people have played in creating the countryside.

WHEN we talk about saving the countryside we are really talking about defending our national heritage. The English landscape is our great collaborative masterpiece and our greatest gift to the wide world, greater even than Shakespeare. The heroes of this book – from Wordsworth and Ruskin to Octavia Hill and Patrick Abercrombie – instinctively understood that because our greatest achievement is to have made the English landscape, our greatest responsibility is to save it.

And placing great responsibility on great minds can so often produce great ideas. These people challenged received wisdoms and succeeded in imprinting genuinely radical ideas in the collective national psyche. Great practical ideas like Green Belts have physically saved so much of our countryside, while less tangible concepts – like Wordsworth's 'Sense Sublime' – have captured our imaginations and helped create a national connection to the countryside, and a natural urge to defend it. And then there are ideas like the Right to Roam, which have encouraged millions of people to interact with the countryside and feel the sense of ownership that is at the root of any action to save it.

And for all the visionary greatness of those who have the initial ideas, it is the willingness of the rest of us to climb the barricades which guarantees their success. In his 1971 account of *The Survival of the English Countryside*, Victor Bonham-Carter ventured that 'the progress of an idea, its origin as a minority fad, its growth and gradual acceptance by the public' relied on two main forces: 'the foresight, devotion and persistence of a handful of men and women at the head of the movement' and, equally, the efforts of thousands of volunteers to engage with the machinery of government and provide local evidence of the need for change.

As President of the Campaign to Protect Rural England – an organisation which relies on its extraordinary local volunteers – I've seen at first hand that people are as passionate about the countryside as ever. But with the current preoccupation with economic growth at all costs, it is no exaggeration to say that our countryside is in greater danger than it has even been. That is why we must reassert the relevance and importance of the time-honoured ideas that have saved it, and can continue to keep it safe.

These 22 Ideas are part of what makes us who we are; national icons as much as the NHS. They are among the things that make England, and Britain, great. They should not be abandoned without a fight.

Introduction

OPPOSITE
Devon's peaceful pastures and bustling hedgerows retain their bucolic charm despite centuries of urbanisation in England.

A book celebrating the ideas that saved the English countryside must take the rather optimistic viewpoint that the countryside has, to some meaningful extent, been saved. Before we can assess whether we really have saved the countryside, we must consider how much countryside we still have, and how much has been lost. The archaeologist Francis Pryor wrote, in *The Making of the British Landscape*, that 'even though nearly everyone lives or works in a town or city, somehow Britain has managed to retain its uncluttered rural areas. We take these things for granted, but I consider them a huge achievement.'

England is still predominantly rural and – as politicians and developers are always happy to point out – only around 5 per cent of England's surface area is built on. Government statistics classify 18.5 per cent of England's land as 'Urban or other (non-agricultural)', although this includes 'non-agricultural semi-natural areas such as grouse moors, sand dunes and grass land'. The official estimate for the area of truly urban land use in England is around 11 per cent, although this percentage does not include open space in urban areas, such as parks and gardens – hence the fact that only 5 per cent of the country is actually covered by concrete, brick or tarmac.

Harry Mount, in *How England Made the English*, wrote that 'to an astonishing degree, much of the countryside retains its pre-war look, despite England being one of the most densely populated countries in Europe'. Recent reports suggested that England is, in fact, *the* most crowded major country in Europe. As a famously 'small island' – and the world's trailblazing industrial nation – England's landscape has had to accommodate rapid growth in its cities and in the industry and infrastructure that sustain them. It is truly remarkable that we can still talk of England as a 'green and pleasant' land without a hint of sarcasm. With over 400 people per square kilometre, it is around twice as crowded as Germany and Italy, four times as crowded as France, and six times more densely populated than our neighbours in Scotland. And yet, when we look out of an aeroplane window, or hover over satellite images, an unmistakably rural scene confronts us.

Of course, statistics never tell the whole story. Part of the surprise people feel when learning how much countryside we have left is caused by their perception of England as highly urbanised. Eighty per cent of us live in towns and cities, and many of those who do not still have

England from above is still a blanket of green, with its cities – even London – merely grey specs in comparison to the sprawl of concrete visible on the continent.

to commute to conurbations. Our daily lives are defined and confined by concrete, steel and glass; most of our journeys are along roads and railways, where development is most likely to be concentrated. Busy lifestyles allow for only fleeting visits to real countryside; and we very rarely find ourselves in genuine wildernesses, away from any of the sights and sounds of urbanisation.

We are an urban society, occupying a miraculously rural country. But, in a way, our perceptions are right – we are not quite as rural as we think we are. We would not call an urban back garden 'countryside'; no matter how peaceful and full of wildlife they can be, gardens are bound by fences and houses. In the same way, our great urban parks can be a wonderful escape, but they are unmistakably urban – often surrounded by railings, roads and skyscrapers. While almost 90 per cent of the country is undeveloped, the urbanised percentage is spread across the whole country, not just as cities, towns and villages, but also as roads, railways, powerlines and quarries. These urbanising elements take up a relatively small amount of land, but by criss-crossing the country they 'intrude' on our perception of rural.

INTRODUCTION

Tranquillity is a huge part of what constitutes our idea of the countryside as an escape route from the sights and sounds of human society; these interventions make it almost impossible to find not just silence, but solace, in natural sounds like running water and birdsong. So while the vast majority of England might be considered rural, research by CPRE found that over 50 per cent of the country suffers from this intrusion.

While acknowledging that we have saved less countryside than we would have wished for, this is not a pessimistic book. There is plenty to celebrate, and in a densely populated country, the control of the growth of the biggest threat to the countryside – our towns and cities – is a huge achievement. England's transition from a rural to urban society was astonishingly rapid. In 1750, aside from London there were only six towns that would have been considered 'urban' by the latest Government definition (over 10,000 in population). Seventy-five per cent of the population lived in rural areas – in hamlets, villages and small market towns – but just a century later the majority of people lived in urban areas.

Energy infrastructure – like these pylons and wind turbines in East Sussex – has a relatively small footprint, but its visual intrusion has changed the face of much of the countryside.

'... much of the countryside retains its pre-war look, despite England being one of the most densely populated countries in Europe.'

Around a million people lived in towns and cities in 1750; by 1901, this had risen to over 23 million, with the biggest engine of growth, London, having grown to occupy over 120 square miles.

Remarkably, after two centuries of fairly organic growth in its physical size, London began to sprawl rapidly without a commensurate growth in population. In the fifty years after 1901, the area of land covered by the capital more than quadrupled to around 500 square miles despite its population only growing by around 30 per cent (peaking at just over 8 million). In 1951, the population of Outer London outnumbered Inner London for the first time. This unprecedented spread of the city was the result of a number of factors: the desire to improve urban living conditions through the kind of low-density suburban development popularised by the first Garden Cities; the building of 1 million council houses between the wars (and a further 2 million by speculative builders with Government subsidies); and the rise of transport options – the car, commuter train and the Underground – which made it possible to chase the dream of working in the city while living near the country.

But the main reason for the explosion of growth in London and England's other towns and cities, was simply because they could. There was no effective barrier against urban sprawl until the Town and Country Planning Act of 1947 and the national introduction of a Green Belt policy in 1955. Thanks to these controls, since 1951 the outward growth of London has returned to a pre-1900 rate, reaching 600 square miles by 2011. Though London's Green Belt has been nibbled at over the years, its actual area continues to increase – at 2,000 square miles, it covers almost 4 per cent of England's total land area – and the runaway physical growth of the capital has been slowed. And this has clearly not been to the detriment of economic and cultural success; London's importance as a global centre has soared since 1951, culminating in the 2012 Olympics.

The Games, on a former industrial site in Stratford, also illustrated how as a nation we have embraced urban regeneration as an alternative to building on green fields. Constraining the outward growth of urban areas has helped to incentivise the re-use of derelict land within cities and to avoid the 'donut effect' seen in Detroit and many other American towns and cities. In fact, a cursory investigation of urbanisation in the States is a glimpse of what might have been in this country. It is estimated that

INTRODUCTION

if London had been allowed to grow at the same rate and density as Los Angeles since the Metropolitan Green Belt was created in 1938, it would now be within touching distance of Brighton and Cambridge.

So we can confidently say that much countryside has been saved for future generations by the pioneering thinkers and campaigners who originated the ideas explored in this book and made them work in reality. We will explore how their once radical ideas were developed, popularised and ingrained in our collective national psyche.

The ideas that shaped the history of countryside conservation in England came in response to the Industrial Revolution, during a ninety-year period of campaigning, beginning in 1865 with the formation of England's first conservation body and ending when Green Belts became national Government policy in 1955. But the scope of this book is much wider than those ninety years. We will look at the inspiration behind those pioneering campaigners, and discover just why they gained such influence in the first half of the twentieth century. We will see that, though many battles were won, the fight for the countryside continues to this day, and we will show how the modern 'environment' movement evolved through the last fifty years.

Salisbury Cathedral from the same spot in the 1920s and 2000s; suburbia claimed huge quantities of countryside before the introduction of Green Belts and a Democratic Planning system. But the losses have slowed dramatically: between 1927 and 1939, 60,000 acres were built on each year, by 2013–14 that had fallen to 7,500 acres.

Cultural phenomena such as primogeniture and enclosure have undoubtedly shaped the English countryside – together with great physical operations like the excavation of the Norfolk Broads by centuries of peat-diggers, and the draining of the Fens by the great Dutch engineer Cornelius Vermuyden. Though the unforeseen consequences of these actions may well have helped save – and even create – a great deal of rural England, we restrict ourselves to the discussion of the ideas that were promoted specifically to protect the countryside; new ways of thinking, which challenged prevailing attitudes and became popular causes within a generation.

Our countryside is a lasting legacy, which could so easily have been lost if not for the visionary ideas of the pioneers celebrated in these pages. England has been uniquely blessed, not just with the ideal location, climate, geology and geography to produce perhaps the greenest and most engaging landscape in the world, but with three successive generations of men and women uniquely endowed with the qualities to save it – campaigners who seemed born to their task, capable of presenting their arguments with prose that was almost poetry.

Idea 1

Discovering the Landscape

introduced by Melvyn Bragg

Nature arrived more like a bang than a whimper. The Industrial Revolution meant Hell for increasing thousands and then millions of industrialised workers in cramped, polluted, neglected cities. Nature became an earthly paradise which offered escape and calm and communion with the rest of life.

Now, most of us live for most of the time in an urban landscape radically improved. Yet the countryside has become something of which we are even more conscious, and we see it as much more precious now that there's less of it.

This country was extremely lucky to have one of its greatest poets, Wordsworth, born in and then living for most of his life in the Lake District, and devoting his life to the poeticising of Nature. Alongside Wordsworth we had the extraordinary detailed nature paintings of Turner and Constable, who brought glory to the ordinary and captured the interest and the imagination of many people who had either never seen a landscape painting or never rated it as a worthy subject comparable to say a historical or religious painting.

Nature was the healer for Wordsworth. As time has gone on it has become that which itself must be healed. Now, so soon after Nature has leapt into our consciousness, we find ourselves not only holding on to it more and more tenaciously but fearing that we might lose it and slowly – too slowly for some – bending our efforts to the preservation of the planet's greatest gift.

IDEA 1 • DISCOVERING THE LANDSCAPE

PAGES 14–15
The beauty of Colmers Hill emerging from the Jurassic Coast mist.

OPPOSITE
Where the view of Buttermere village from Red Pike might once have induced feelings of 'horror and immensity', our changing relationship with the landscape now allows us to appreciate its intimate beauty.

Before we could even think about the notion of saving the countryside, we had to learn to see it with new eyes. Until then, nature and landscape would continue to be seen either as the mundane backdrop to a life of toil or as a terrifying wilderness of the dangerous and unknown.

For most of the history of humanity, the look of the landscape was largely taken for granted. When Eugenius reminded the Emperor Constantius I that the Roman Empire could not afford to lose the land of Britain, it was for purely practical reasons: 'So plentiful are its harvests, so numerous the pasture lands in which it rejoices.' Throughout the Middle Ages the natural world was seen as a gift from God for humans to exploit; as the historian Keith Thomas puts it, 'man's ascendancy over the natural world was the unquestioned object of human endeavour'. The land was a resource, and beauty was not a quality associated with it; there was no conception of the notion of a 'view'. The writer Adam Nicolson has described medieval vision as being 'intensely particularised', focusing on the detail of the individual elements that were relevant at the time – the animals being hunted, the soil being tilled, or the fruit being picked.

In the sixteenth century, most people in England would have been utterly, almost mind-numbingly familiar with the verdant open spaces that today we are forced to seek out. Even in the largest towns of the time, you were never more than a ten-minute walk from open countryside. And for most, the strongest association with the natural world would have been one not of enjoyment, but of long hours of back-breaking labour under unforgiving skies.

RECORDING OUR SURROUNDINGS

In the medieval period, landscape only permeated popular consciousness as a backdrop for religious painting, or, in the upper echelons of society, as an illuminated record of profitable land holdings. By 1520, Joachim Patinir had established himself as the first Dutch master of landscape, pioneering the idea that paintings might be enjoyed purely for the attractiveness of the natural scenes they portrayed.

Patinir's landscapes leaned towards fantasy interpretations, but in 1565 Pieter Bruegel the Elder's paintings of the seasons gave a much more realistic interpretation of contemporary countryside. For the first time, Biblical interpretation is replaced by an attempt to capture the power of nature. Bruegel's earthy and unsentimental observations reflect what would then have been a seismic shift in the way artists represented the world around them.

The work of Patinir and Bruegel was so revolutionary that it needed a new word to describe it. In 1520, Patinir's contemporary Albrecht Dürer had described him as a good painter of *landschaft*, a German word meaning the shape or condition of an area of land. 'Landschap' (soon evolving to 'landskip') became the common form used in the Netherlands to describe the painted surveys the aristocracy

The Camp on Warley Common (1778) by Thomas and Paul Sandby showed painters beginning to put the English landscape centre-stage. This trend could be said to have been started by Gainsborough's 1750 portrait of *Mr and Mrs Andrews*, which also devoted over half of the canvas to a panoramic view of the Essex countryside.

commissioned as a visual record of their landholdings. As the painters took complete ownership of the word, it became all about the view *of* the landscape, rather than the physical state of things *within* the landscape.

APPRECIATING THE VIEW

In 1600, the English countryside was something one had to escape *from* – usually to find civilisation in the city. Travel writers conducting 'grand tours' of the country were far more likely to record urban pleasures than offer any compliments to rural scenery: from John Leland on the 'beauty' of Birmingham in 1530 to Celia Fiennes taking pleasure in 'neat' towns in 1690. By the mid-eighteenth century, the increasingly smelly, noisy and overcrowded towns meant that the country was becoming very attractive by comparison.

The appreciation of natural landscapes, both wild and 'man-made', depended on some knowledge of the work of Claude Lorrain – one of the first artists, around 1629, to paint a landscape without figures for its own natural beauty. English sensibilities were educated and tastes primed by the likes of Claude, to the point that by 1680, most fashionable middle-class homes were embellished by a Dutch or Italian 'landskip' print hanging from the wall. Over the next century, English artists inevitably tried to emulate their continental cousins; in 1774, Thomas Gainsborough complained of being 'sick of Portraits' and wished 'very much to walk off to some sweet Village where I can paint Landskips'.

Paul Sandby had come from topographical work for the fledgling Ordnance Survey, but is perhaps best known for his vast panoramas of the 1760s, commissioned by the aristocracy to show off their property. Sandby had a tendency to fill the foreground with the less celebrated parts of these estates, rather than the expensively landscaped areas the owners were most proud of. His brother and partner, Thomas Sandby, used a Royal Academy lecture of 1775 to explain that although the formal gardens of these estates were private, 'all the surrounding country within our view may be looked upon as our own property, when considered with regard to pleasure . . . as if placed there by our own expense'. When George III commissioned the Sandbys to paint the British Army assembled at Warley Common in 1778, he wanted to capture what the visiting Samuel Johnson described as 'one of the great scenes of human life'; what they gave him was a majestic panorama of the Essex countryside with the soldiers barely visible in the distance. While it is not known how the King reacted, his later comment that Paul Sandby could 'turn his hand to anything, like a fox' has been interpreted as not entirely complimentary.

THE FIRST ENGLISH TOURISTS

George III would have more reason to be pleased with Sandby when engravings of his paintings of Windsor Great Park began appearing in new media like *The Copper Plate Magazine* at this time. The popularity of these views were priceless propaganda for George; their growing familiarity gave the nation a sense of ownership over these landscapes, and this patriotism was only enhanced when Sandby portrayed the felling of Windsor's trees to supply the Napoleonic war effort. Before the wars, Horace Walpole had

In the mid-eighteenth century Kent's 'homely and familiar' beauty was criminally overlooked by the landscape connoisseurs of the Grand Tour. By the 1890s, William Morris was arguing that 'the lead of our ordinary English landscape becomes pure gold' when we recognise its 'character'.

'As Grand Tours of Europe became increasingly dangerous and impractical, the thirst for views had to be satisfied domestically.'

lamented, in 1764's *The Castle of Otranto*, that England's 'ever-verdant lawns, rich vales, fields of hay-cocks, and hop-grounds, are neglected as homely and familiar objects'. The conflict made the appreciation of English landscape not just an act of patriotism but a necessity; as Grand Tours of Europe became increasingly dangerous and impractical, the thirst for views had to be satisfied domestically.

While the Sandbys helped make English landscapes popular, the fact that people were starting to visit them for pleasure was largely thanks to Reverend William Gilpin's promotion of the 'picturesque'. Gilpin coined the term in 1768 for views which had 'their own particular beauty which would be agreeable in a picture', helping to create a boom in rural sightseeing which was to fill the void left by the inaccessibility of the European landscapes made fashionable by Claude.

Perhaps Gilpin's greatest achievement was to ensure that the users of his travel guides actively engaged with the landscape, by encouraging them to make sketches and pay attention to the details of the countryside. However, Gilpin was often unsure of the aesthetic merits of the places he urged people to visit, bemoaning the 'rudeness' and 'deformities' of nature, which a 'practised eye would wish to correct'. Gainsborough also doubted whether a depiction of 'real views from nature in this country could afford a subject equal to the poorest imitations of Claude'. Rather than an outright slur on the English countryside, this may well have been an admission that realistic landscape painting would always struggle to compete with the idealised forms an artist could conjure from 'his own brain'.

Gilpin's complaints were a hangover from the travelogues of the turn of the eighteenth century; Celia Fiennes' verdict was that the north of England was full of 'inaccessible high rocky barren hills which hang over one's head in some places and appear very terrible'. Daniel Defoe described the north as the 'wildest, most barren and frightful of any place'; where the Lake District began, he felt the 'pleasant part of England was at an end'. The historian G.M. Trevelyan later put it more viscerally, suggesting 'the fear of having one's throat cut at the next turn of the track was not conducive to picturesque raptures amid Highland scenery'.

Possibly the first person to challenge this interpretation was the artist who became Gilpin's tutor. In 1753, Dr John Brown acknowledged the 'Horror and Immensity' of the Lake District, but felt it was tempered by its beauty: 'delicate sunshine over the cultivated vales' and 'the majesty of

The classic view from Symonds Yat in Herefordshire was the birthplace of Gilpin's Picturesque in 1782, when his *Observations on the River Wye* created the idea of a guided tour based around fixed viewpoints.

the impending mountains'. Within a century of Celia Fiennes describing the Lake District as 'desert and barren', there had been a remarkable surge in the attraction of its 'wild' nature. Visitors to the Lakes increased hugely after 1773 when a regular coach service from London combined with a service taking explorers over Shap Fell.

The change in taste can partly be explained by the rise of large-scale landscape gardening between 1740 and 1760. Encouraged by the growing visual appreciation of their patrons, the likes of Lancelot 'Capability' Brown began to create informal, curved lines in parks around great houses to contrast with the increasingly geometrical layout of surrounding fields as a result of enclosure. Where, once, formal gardens had been created as a respite from 'wild' countryside, Trevelyan argued that Brown's popularity was a response to 'the fact that nature was getting a little too tame'.

CAPTURING THE TRUTH OF ENGLAND'S BEAUTY

The challenging ideas of Gilpin, the Browns and the Sandbys paved the way for the great celebration of the English landscape by John Constable. Declining to 'seek the truth at second hand', Constable focused on uniquely English landscapes and in particular, those he was intimately acquainted with: 'The sound of water escaping from mill dams; old rotten planks, shiny posts and brickwork; these scenes made me a painter, and I am grateful.'

Even though he was a devotee of Claude, Constable was happy to eschew the Grand Tour and embrace his role as England's unofficial national painter. Kenneth Clark noted: 'We have got so used to this approach to painting that it is difficult for us to see how strange it was ... at a time when all serious artists aspired to go to Rome.' Constable himself said: 'I was born to paint a happier land, my own dear England – and when I forsake that, or cease to love my country – may I as Wordsworth says "never more, hear her green leaves russel or her torrents roar".' Although it is Turner who is more closely associated with Wordsworth, Clark argued that Wordsworth's great kinship was with Constable: 'They both grasped nature with the same physical passion. Both loved their own places.'

Constable's loyalty to the detail of ordinary England did not endear him to the art establishment. The Royal Academy managed to exclude him until a year before his death and his influence was felt more strongly away

A dung heap might seem an odd subject matter for a wedding present, even if the bride had grown up near this view of the *Stour Valley and Dedham Church*. But Constable could not resist the earthy charm of rural England, especially in 1814 when he did most of his painting directly from nature, avoiding the temptation to embellish the truth in the studio.

The setting of Constable's *The Hay Wain* as seen today; his paintings captured a country on the cusp on industrialisation, and helped create the will to save the 'everyday' beauty of England.

from home, being particularly admired in a post-revolutionary France craving the calm certainty of his landscapes. Although Constable's brushwork was often radically modern, his subjects were deemed to be too ordinary. Patrons wanted to be transported by paintings, not confronted with reality – especially if it was something they could see on their own estate. But what made Constable unfashionable in his lifetime has made him hugely popular ever since. He captured England on the cusp of industrialisation in a way that no other painter attempted. Constable froze those landscapes in time, and that is why we automatically think of his work when we try to conjure up the pastoral idyll. He connects us to our roots, documenting the last years of England as a predominantly rural society.

Constable's relationship with printmaker David Lucas also ensured the growth of his popularity. Prints helped landscape become a more affordable genre and Constable's series 'Various Subjects of Landscape, Characteristic of English Scenery' (1830–32) capitalised on the public desire to reminisce and to celebrate the views and ways of life that were rapidly disappearing by the mid-nineteenth century. Because of this, Constable has always been associated with nostalgia – and often derided as a 'chocolate box' painter. But it was Constable who did more than anyone to make the English realise that their ordinary countryside was beautiful and worth protecting. Dedham Vale was nothing, was ignored, until Constable painted *The Hay Wain*.

In 1894, fifty-seven years after Constable's death, William Morris asked why 'we refuse to pay attention to anything in nature which is not tremendous and exciting'. The architect Clough Williams-Ellis continued this theme in 1928, arguing: '"The beauty about us" – that is the beauty of country, town and village, the normal visible setting of our ordinary everyday lives.' While Wordsworth's championing of our most sublime upland landscapes has received most recognition from conservationists and the environment movement, these places never faced the same level of threat as the less spectacular countryside around and between our towns and cities. That is why it is Constable's articulation of the beauties of the ordinary countryside that has become part of our shared national psyche, and why Government policy now officially recognises the 'intrinsic character and beauty of all countryside'.

Idea 2

Preservation

introduced by John le Carré

I wish it were so simple. But it isn't any more. Against the great, crass violations of our landscape – if we have the vigilance, the spirit and the support – we can raise a storm, pester planners, constructors, councils, ministries and tribunals, and the law courts. And now and then we may win. But against the endless drip-drip of smaller violations that over time will cause just as much damage to our beloved countryside, we seem almost powerless. In this war of attrition the enemy is everywhere. He is the corrupt council official in search of informal reward. He is the unscrupulous property dealer with an army of lawyers who gradually wears down an overstretched planning committee with repeated special applications, until all of a sudden a cluster of 'affordable' houses, which are anything but affordable, pop up on a pristine hill slope 20 miles from the nearest place of work. He is the cash-strapped dairy farmer toiling in a designated place of outstanding natural beauty, who must build that 'essential' bungalow in order that a cowman who doesn't exist can live close to cattle he doesn't tend. He is the faceless entrepreneur who lobbies in high places for his right to erect wind turbines, mobile phone masts and listening masts that will desecrate yet another piece of unspoiled land in the supposed interest of the public. He is the politician who, in the catch-all name of free enterprise, unlocks the last restraint on our rush to destroy the most precious heritage that we share: our vanishing English countryside.

By the time we accepted that the English countryside was worth protecting, it was under serious threat from the forces of industrialisation, compounded by the prevalent notion that *laissez faire* government should allow private individuals to do what they liked with their land. William Blake famously broached the tension between the rise of the 'dark satanic mills' and the survival of 'England's green and pleasant land' in his introduction to *Milton* in 1808. The following decade saw artists like John Linnell and George Lewis begin to document the westward spread of London over the surrounding countryside, a theme pioneered by Thomas Jones' 1785 illustration of countryside on *The Outskirts of London* being turned into a building site.

By 1814, Wordsworth could give a sense of the speed of the changes engulfing the countryside, documenting the impact of the urban sprawl created by the industrial revolution. In *The Excursion*, Wordsworth remarked (probably of Manchester):

> Meanwhile, at social Industry's command,
> How quick, how vast an increase! From the germ
> Of some poor Hamlet, rapidly produced
> Here a huge Town, continuous and compact,
> Hiding the face of the earth for leagues . . .

Fifteen years later, George Cruikshank's cartoon 'London Going out of Town or the March of Bricks and Mortar' satirised the speculative development of the green fields of Islington. In 1841's *The Old Curiosity Shop*, Charles Dickens described the sprawl of Birmingham into Wolverhampton as 'a long, flat, straggling suburb . . . a cheerless region, where not a blade of grass was seen to grow'. But behind the veneer of satire lay a real sense of panic. Towards the end of his life in 1841, Wordsworth reacted to the expansion of the railways into the Lake District by asking: 'Is then no nook of English ground secure from rash assault?'

Wordsworth's natural successor was the art critic John Ruskin, who in 1859 warned of the dangers of turning 'the south of England into a brickfield, as we have already turned nearly the whole of the north into a coal-pit'. Ruskin feared that 'changes in the state of this country are now so rapid' that 'no acre of English ground shall be without its shaft and its engine; and therefore, no spot of English ground left, on which it shall be possible to stand, without a definite and calculable chance of being blown off it, at any moment, into small pieces.' The pace of industrialisation meant that preservation was becoming an urgent measure. The Victorian zeal for England's culture and heritage fostered a growing sense that this generation had a moral duty to preserve, and no more keenly was this duty felt than in the House of Commons.

PAGES 24–5
The Newlands Valley, saved from a slate railway following one of the earliest campaigns for the preservation of the Lake District in 1883.

IDEA 2 • PRESERVATION

LONDON going out of Town. — or — The March of Bricks & Mortar!

By 1829, the loss of London's surrounding fields was a topic ripe for the satire of George Cruikshank. But bricks and mortar weren't the only threat to the countryside: the following year saw the economist John Stuart Mill oppose a railway through the Vale of Mickleham in Surrey he felt was 'unrivalled in the world for exquisiteness'.

DEFENDING THE COMMONS

England's first environmental campaign group was formed by an MP, George Shaw-Lefevre (later Lord Eversley), who in 1865 founded the Commons Preservation Society (CPS) with support from John Stuart Mill and Ruskin's protégé, the poet and decorative artist William Morris. By this time, the rural enclosure movement had reached the edge of England's towns and cities, where ancient rights to land were being reclaimed by landowners for a more profitable use: not farming – which the enclosure movement was originally intended to promote – but real estate. Of course, the commons were increasingly valued as an in-built check on the spread of cities, and a place of recreation for those who lived in them.

Shaw-Lefevre had been closely involved in two years of Parliamentary debates about the threat of development of Hampstead Heath, Epping Forest and Wimbledon Common. Today, we think of these places as urban oases, but in the mid-nineteenth century they were London's nearest countryside; the preservation of commons was effectively concerned with the restriction of urban sprawl, rather than maintaining urban green spaces. In February 1865, Shaw-Lefevre proposed a Select Committee to inquire into the best means of 'preserving for the public use the forests, commons, and open spaces in and around the metropolis'; he stated that there were 'about seventy commons and greens of various sizes within fifteen miles of London, which were all in jeopardy under the present state of things'.

In the same debate, Henry Cowper MP suggested that such a wealthy city as London should be able to buy and preserve these spaces for the people, noting that 'Saffron Hill, Rosemary Lane, Mayfair, and other places whose pleasant names recalled the memory of agreeable walks, had long since been engulfed' because of the 'temptation to convert the rights in commons into money'. MPs like Cowper and Shaw-Lefevre were worried that, while very few questioned their arguments in favour of preservation, Parliament's reluctance to 'confiscate private property' would lead to a 'process of extinction' for the commons that 'belted' London with 'natural parks'.

Lady's Walk leading to Berkhamsted Common, the scene of one of the most remarkable successes of George Shaw-Lefevre's Commons Preservation Society.

THE CASE FOR LEGISLATION

In March 1866, Cowper introduced a Bill intended to prevent London from engulfing the 'natural and unpretending charm' of its 'circle of fresh and breezy commons'. Cowper made a passionate case for the preservation of the 180 commons within a 15-mile radius of the city – an area of land totalling 10,500 acres. His speech would be an early example of what would become a defining characteristic of the preservationist campaigners – the ability to put their arguments across with the panache of poets; Cowper said of the commons:

> No institution handed down to us from our Saxon forefathers had contributed more to the happiness of the people. As the great Babel is encroaching on the country – just as those commons are becoming more valuable for the enjoyment and recreation of the people, they are being more menaced by the builder and the railway engineer.

Sir Francis Powell seconded Cowper's motion that the commons must be kept open to secure 'the blessings of fresh air; that the next generation, instead of being weaker, might be more powerful'. At its second reading, Cowper argued that the Bill was a response to the 'strongly expressed wish of the public for some legislation to prevent the destruction now so frequently observed of places of natural beauty, health, and enjoyment around the metropolis, which were being encroached upon by buildings and railways'. The Tower Hamlets MP, Acton Ayrton, made the case for preserving commons for the Londoners who 'have to pass over miles of roads before they can find a green shrub or tree uncontaminated by the smutty nature of the London atmosphere'.

The Metropolitan Commons Act was passed in August 1866, giving Inclosure Commissioners powers to prevent the enclosure of land within 15 miles of London where previously, they had only been able to refer these cases to Parliament. The new legislation challenged a loophole that had allowed the well-meaning Inclosure Acts to, in Cowper's words, 'become a public evil'. Designed to increase agricultural employment and food production, the Inclosure Acts has an unintended consequence: 'to increase the possessions of the lords of the manor' by allowing fields to be enclosed for building on.

But even before the new Act was passed, Shaw-Lefevre and his CPS colleagues had organised a trainload of navvies to help pull down new fences which had illegally enclosed Berkhamsted Common. The Common remains to this day and the new legislation helped the CPS go on to save Hampstead Heath from gravel extraction and keep Wimbledon Common and Epping Forest open for public enjoyment.

'The external aspect of the country belongs to the whole public, and whoever wilfully injures that property is a public enemy.'

TOWARDS A NATIONAL MOVEMENT

Despite these successes, the power of the landowners meant that the Metropolitan Commons Act could not completely halt the lucrative developments of the commons around London. The Lords of the Manor could use Private Bills to bypass the protection of the Inclosure Commissioners, while the Act did not provide any extra funds if local authorities could not afford to fully compensate landowners for the loss of development value. Their inability to purchase common land in such cases would ultimately encourage the CPS lawyer Robert Hunter to explore the ideas that would lead to the creation of the National Trust.

The loss of commons tended to go hand in hand with the loss of footpaths, and the CPS joined forces with the fifteen-year-old National Footpaths Preservation Society in 1899 to create the Commons and Footpaths Preservation Society. Footpath preservation groups had been formed in Manchester and York as early as the 1820s, and in 1833, the report of the Select Committee on Public Walks recognised the need to provide paths and open spaces for 'comfort, health and content'. Robert Slaney MP described the growing need for footpaths by highlighting that although 'the population of Manchester is 187,000, there is not an open suitable place for exercise or recreation'. On the rural fringes of the town, working people attempting to get into the countryside were 'met on the road with notices against trespass, and the inhospitable intimation of spring-guns, and Steel-traps'. Slaney also noted that the moral, as well as physical, well-being of the working man was at stake, with 'the poor workman forced into the public house' in the absence of more wholesome amusement.

The growing influence of the early campaigns to save footpaths and commons gave rise to a rapid increase in single-issue pressure groups seeking to preserve very distinct aspects of our natural and built heritage. First ancient buildings, then monuments, metropolitan public gardens and birds – all had their own national body fighting for their interests within twenty-five years of the formation of the Commons Preservation Society. This period of 'preservation mania' represented the outpouring of public support from those who now shared the early concerns of Wordsworth, and reflected the growing influence of the dominant voices of the preservation movement from the mid-nineteenth century: Ruskin and Morris.

IDEA 2 • PRESERVATION

William Wyld's view of *Manchester from Kersal Moor* of 1852 shows why campaigning MPs like Robert Slaney were so keen for its workers – who Queen Victoria (for whom the painting was commissioned) described as 'painfully unhealthy-looking' – to be able to escape into the surrounding countryside.

DECLARING WAR ON INDUSTRIALISATION

An early supporter of the Commons Preservation Society, Morris crystallised the moral certainty of those early campaigns in an 1881 speech entitled 'Art and the Beauty of the Earth'. Morris realised that in order to grow and succeed, the preservationists had to convince people that the countryside was theirs to enjoy, but that a powerful enemy wanted to deny them that right: 'When you have accepted the maxim that the external aspect of the country belongs to the whole public, and that whoever wilfully injures that property is a public enemy, the cause will be on its way to victory.' Despite this optimism, Morris was clear that the enemy was growing ever more powerful:

> We must turn this land from the grimy back-yard of a workshop into a garden. . . . Time was when it was beautiful from end to end, and now you have to pick your way carefully to avoid coming across blotches of hideousness which are a disgrace to human nature. I have seen no statistics of the size of these blotches, but in some places they run together so as to cover a whole county, or even several counties, while they increase at a fearful rate.

Morris expressed the fear that the scale of the task had created apathy among those who regarded the challenge as 'difficult, or rather impossible'. However, he was greatly encouraged by the rapid growth of the preservation movement: 'If I am a crazy dreamer, there are many

Rockford Common in the New Forest, where common rights have allowed the survival of the ancient practice of 'commoning' – a system of farming dating back to William the Conqueror – and helped shape the unique beauty of the landscape.

members and supporters of such societies as the Commons Preservation Societies, who have not time to dream, and whose craziness, if that befell them, would be speedily felt throughout the country.' With his ideas of a society based on harmony between man, art and nature, John Ruskin had been a mentor to Morris and many others in the early years of the preservation movement. By the 1880s, Ruskin – then in his sixties – was actively helping to speed the spread of Morris' 'crazy dream'.

In 1883, Ruskin joined forces with Robert Hunter, his former Oxford student Canon Hardwicke Rawnsley, and Octavia Hill, who had been campaigning with the Kryle Society to 'Bring Beauty Home to the Poor' by defending urban heaths and woodland. Together, their campaign opposed the construction of railways to carry slate through the unspoilt Lake District valleys of Newlands and Ennerdale. The action placed England's most spectacular landscape at the heart of a preservation movement that hitherto had been focused on saving relatively ordinary places on the cusp of urban sprawl. The implication was clear; if the world-renowned Lakeland scenery could be compromised, nothing was sacrosanct. The subsequent formation of the Lake District Defence Society attracted patronage from the poets Robert Browning and Alfred, Lord Tennyson, as well as London's greatest landowner, the Duke of Westminster.

A NEW MANIFESTO FOR PRESERVATION

After the Lakes, the most celebrated place in England was the Hampshire village of Selborne, with Gilbert White's book on its *Natural History and Antiquities* having gone through seventy editions since his death in 1793. In 1885 a Devonian couple, George and Theresa Musgrave, set out to promote White's ideas through their Selborne League for the 'preservation of forests and places of popular resort by means of publishing any threatened destruction of them'. The preservation of 'birds of beautiful plumage' was a *cause célèbre* of the 1880s, and the Selborne League soon amalgamated with the Plumage League to become the Selborne Society for the Preservation of Birds, Plants and Pleasant Places. With Tennyson as its President, this original incarnation of the Society was a fully fledged national organisation with an office in London, regional branches and subscribing members. Preservation was becoming a highly organised affair.

The impact of the early preservationist movement is perhaps best summed up by a *Spectator* appeal for the Commons and Footpaths Preservation Society of May 1913:

> How great our debt is, and how deep our obligation to pay what we can of it, we may realise a little by trying to picture England as it would have been to-day if the Society had never come into existence.... Imagine London as a city with Hampstead Heath, Wimbledon Common, and Epping Forest partly or wholly built over. Imagine the difficulties in which we should find ourselves in the country in these days when field after field is being caught up in the grip of the town, if there were no Society strong enough to make any private person or local authority think twice before deciding to oppose it.

The preservationists of the second half of the nineteenth century defended many of the green spaces and landscapes we still enjoy today; by the 150th anniversary of the CPS (now the Open Spaces Society) in 2015, over 2,000 square miles of common land survived in England and Wales, an area the size of Lincolnshire. And yet, despite all their efforts, a lack of effective legislation meant that by the First World War much of the countryside was more at risk than ever. At least, as *The Spectator* made clear, developers now had formidable opposition from energetic campaigners, who could mobilise support through simple and heart-felt messages.

Idea 3

The National Trust

introduced by Simon Jenkins

The National Trust was England's 'noblest nationalisation'. It met a clear public desire to save endangered national heritage, whether of buildings or of the natural landscape. This had to involve a degree of statutory control, yet it also had to avoid the statist standardisation to which this might lead. The use of tax relief rather than grant, the retention of former owners as tenants and a firm emphasis on conservation honoured these often-conflicting ambitions. This equilibrium – between 'art and anoraks' – has characterised the Trust to this day. Four million members says it works.

THE formation of the National Trust in 1895 was based on the logic that the best way to preserve a place of historic interest or natural beauty was to buy it and then hold it in trust. The idea originated in an 1884 speech by the then Vice President of the Commons Preservation Society, Robert Hunter, whose experiences with the Society had convinced him of the need for a land-owning 'company' to be able to buy threatened land, not for profit, but to keep it open 'for the benefit of the nation'. When Octavia Hill suggested the company should have a name that suggested benevolence rather than business, such as the Commons and Gardens Trust, Hunter went even further, pencilling simply, National Trust, at the top of Hill's letter.

In acknowledging that land acquired by the Trust would be bought and paid for, Hunter made it clear that his land company would not be formed 'for the spread of political principles' rooted in the aim of the Chartists to make all men eligible to vote by making them landowners. By the time of Hunter's speech, the slogan that every citizen should have 'three acres and a cow' had been adopted by Joseph Chamberlain as part of the Liberal Party's successful attempts to appeal to newly enfranchised rural labourers in the 1885 election. Needless to say, the landowning aristocracy feared such State-led redistribution was too short a step to Karl Marx's recent arguments that only nationalisation of the land could prevent the 'individual abuse' of it. Hunter had to tread a fine line between the socialism inherent in the idea of a National Trust, while appeasing the concerns of landowners the Trust would rely on for gifts of land until they were able to raise funds to purchase it at its development value.

It was not until the beginning of 1895 that the newly knighted Sir Robert Hunter – together with Octavia Hill and Canon Hardwicke Rawnsley – founded the National Trust for Places of Historic Interest and Natural Beauty in England and Wales. Their stated aim was to set aside the best and most beautiful parts of Britain for the public and posterity, and in Octavia Hill's words, 'to provide open-air living rooms for the poor'. The eleven-year delay between Hunter's landmark speech and the formation of the National Trust may be partly attributed to the English establishment's distrust of the Liberal 'intelligentsia', of which the Trust's founders were leading lights. Traditional conservatives would have been genuinely shaken by the notion of private land being acquired for the benefit of the common man. Such a move seemed like the first step towards revolution, especially after February 1886 when, barely eighteen months after Hunter's speech, Trafalgar Square was the scene of violent rioting stoked by the revolutionary Social Democratic Federation.

As fear of revolution subsided it was replaced by growing concern over the consequences of urbanisation. Thanks to the efforts of the preservationists, enlightened landowners were becoming more understanding of the need to preserve parts of their estates. After 1895, the Trust forged good relationships with landowners who would either allow it to manage threatened places, or bequest ownership of the land. The Trust was able to remain independent

PAGES 34–5
Cheddar Gorge in Somerset became one of the National Trust's earliest acquisitions, saving it from quarrying operations.

Toys Hill in Kent was one of the first pieces of English countryside acquired by the National Trust, with Richardson Evan's donation of 1898 soon added to by property purchased by their founder Octavia Hill.

from the State and public donations, while landowners gained good publicity from the high profile donations – it was the perfect late-Victorian good deed.

THE FIRST HISTORIC ACQUISITIONS

Although best known today for its country houses, the Trust's early focus was squarely on protecting open spaces, starting with the gift of 4 acres of Welsh clifftops known as Dinas Oleu by the pioneering philanthropist landowner Fanny Talbot. Twenty years earlier, Fanny had given 4½ acres of land at Barmouth to John Ruskin's 'Guild of St George', a land co-operative dedicated to keeping 'the fields of England green and her cheeks red' by allowing people to make a living from farming and crafts outside the free market system.

As well as taking gifts, the fledgling Trust started to mount publicity campaigns to secure the funds to buy historic landscapes such as 'King Arthur's Cove' at Barras Head in Cornwall, and Cheddar Gorge, which was at risk from quarrying operations. The relatively slow start to land acquisitions raised the question of whether they needed 'the power of the State to protect the antiquity and natural beauties of the country', though Hunter remained hopeful that the Trust 'would eventually become a national institution'. It had already begun lobbying in Westminster,

Derwentwater in the Lake District became the subject of the National Trust's first public major public appeal – to secure the Brandlehow Park Estate – in 1902.

ensuring a clause was inserted into the 1896 Light Railways Bill 'affording a practical means of opposing injurious lines', thereby establishing the 'principle that beautiful scenery was to be respected by railway promoters'.

The slow progress of acquiring land was highlighted by the 1898 AGM's announcement of just two acquisitions, one of which was a piece of land on the spur of Toys Hill donated by the preservationist campaigner Richardson Evans in memory of Mr Frederick Feeney. Giving uninterrupted views of the North Downs, the bequest was announced as 'the first realisation of the idea suggested by the Trust that memorials should sometimes take the form of land commanding beautiful views dedicated to the memory of the dead'. In 1899, the Trust rescued Ide Hill in Kent, and Wicken Fen in Cambridgeshire, but the 1900 AGM acknowledged its 'weapons were only those of sweetness and light' while its total revenue of £330 was 'wholly inadequate to the purpose of acquiring property'. The following year, the Trust acquired only a remnant of a sanctuary cross near Ripon in Yorkshire, with even Sir Robert Hunter bemoaning the inactivity. The Liberal MP Viscount Bryce suggested that State assistance might be required, noting that 'while our scenery is decreasing our taste for it is increasing'.

'While our scenery is decreasing, our taste for it is increasing.'

The State action to invigorate the Trust came in 1907, when it was dissolved and reincorporated as a statutory body by the National Trust Act. The Act enshrined the principles of public access and protection against sale or defacement of land, meaning that the Trust's holdings of 2,000 acres and twelve buildings were now inalienably the possession of the nation. The Trust went from a voluntary outfit that could have been wound up at any moment to what Hunter felt was a body 'authorised and empowered by Parliament'. *The Times* remarked that the new powers would 'furnish a great inducement to donors, who for the first time will be able to feel quite certain of the permanent possession by the public of everything worth preserving that the Trust is able to secure'.

CONCENTRATING ON THE COUNTRYSIDE

The 1907 Act confirmed that the holding of land in trust could be a serious means of saving England's threatened countryside. In the same year, the protracted attempts to acquire Barrington Court in Somerset placed such a financial strain upon the Trust that it would become the last country house it purchased for over thirty years. This allowed the Trust to concentrate on its guardianship of what *The Times* called 'the areas of remarkable natural beauty' and 'still surviving specimens of primeval landscape'. A decade of steady growth was interrupted by the First World War, but in 1923 the Trust made a headline acquisition of a group of Lake District peaks, donated by the Lake District Fell and Rock Climbing Club to commemorate members who had perished in the war. The club's President, Arthur Wakefield, a member of the 1922 Everest expedition, said:

> We have gone back to the grimy cities and to the stale, flat and unprofitable plains inspirited by these mountains to carry on under whatever drudgery we have had to meet. We have stood gazing enthralled at their wondrous beauty and grandeur, and now these mountains are ours and our children's forever.

Accepting the land on behalf of the Trust, the Liberal MP Francis Acland called it 'the nucleus of a great national park. It is a fine thing to know that no future British Mussolini will be able to use Great Gable as a target for his big guns, and that no one will be able to shoot at the Needles with a rifle – though I might like a few more footholds put in it myself! At any rate, no one will be able to shout from below, "come off it".'

The Lake District had been a totem of England's uncultivated wilderness since 1902, when the Trust's first major public appeal – and first foray into the north – succeeded in raising funds for the 100-acre Brandlehow Park Estate in Derwentwater, chosen partly for its accessibility by rail from Lancashire industrial towns. 'Brandlehow will not be preserved for the solitary walker alone, but for the whole travelling public.' Two years later, *The Times* urged the public to help the Trust seize the opportunity to acquire the 750-acre Gowbarrow Park, so that 'Ullswater will in effect be nationalised', and speculative builders denied the chance to 'rob the public of its enjoyment'. Donations for these Lakeland acquisitions came from rich and poor, with factory workers in the North and Midlands among the targets of the appeals. One Sheffield factory worker sent in a small donation to Brandlehow with a note saying: 'All my life I have longed to see the Lakes. I shall never see them now, but I should like to help keep them for others.'

The first thirty years of the Trust had shown that there was a huge public appetite to save beautiful countryside, securing some of our finest landscapes. But, with the growing threats outstripping the pace at which land could be acquired, it became increasingly obvious that an acquisitions policy focused on the most celebrated landscapes would struggle to safeguard the parts of the countryside that were most at risk from development.

ABOVE
Wicken Fen, an iconic fenland landscape saved by the National Trust in 1899; the famous windpump was later donated by CPRE Cambridgeshire.

OPPOSITE
Gowbarrow Park – saved from development in 1904 so that generations of country lovers could enjoy spectacular views of Ullswater.

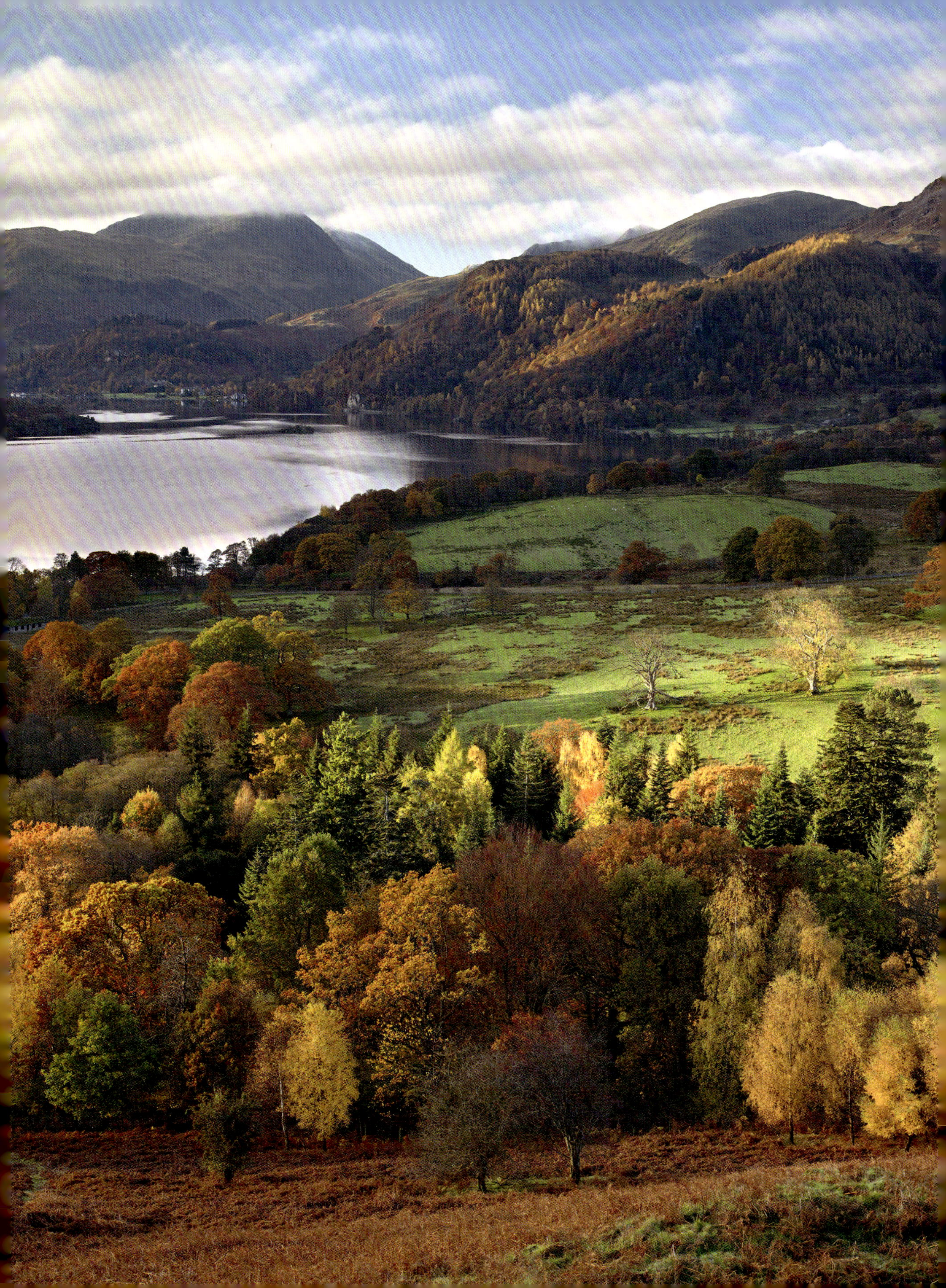

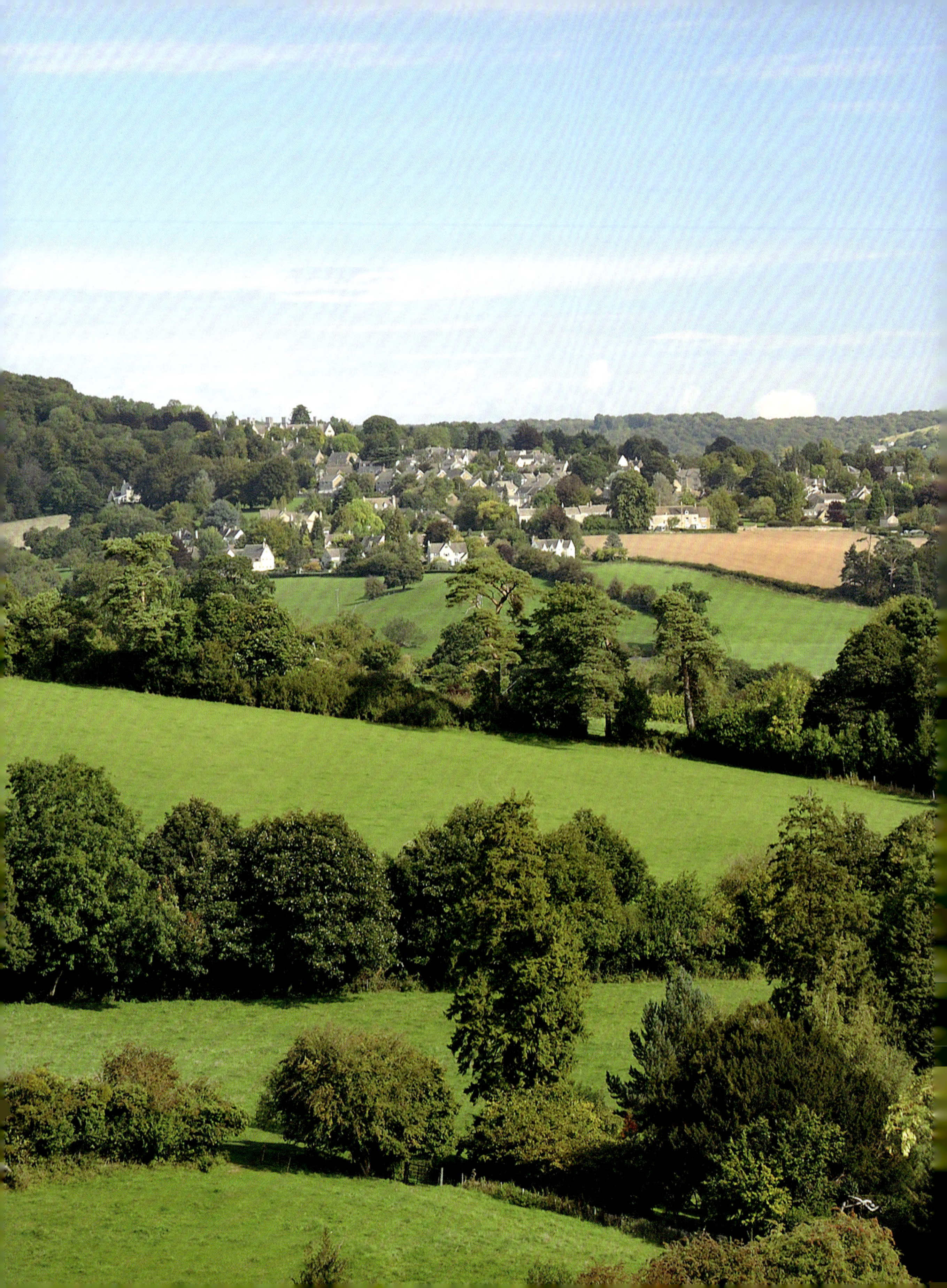

Idea 4

Creating CPRE
introduced by David Puttnam

CPRE's founders had a unique vision for England: a future where new development would allow the economy to grow, and society to progress, without sacrificing the countryside. Effective conservation has always been grounded in the realities of the times, and Patrick Abercrombie's manifesto for *The Preservation of Rural England* made a huge breakthrough by rejecting the idea that a rapidly modernising country had to be 'sterilised or stabilised in its present state' to save its rural heritage.

Abercrombie believed public education and coordinated campaigning – orchestrated by a 'council' of hitherto disparate and highly specialised preservation societies – could encourage the good design and sensitive siting of development. In this way, the whole country could enjoy the benefits of better housing, energy supply and transport without destroying the countryside that Abercrombie felt was 'the greatest historical monument that we possess, the most essential thing which is England'.

Ninety years on, with so much of Abercrombie's countryside remaining – in spite of a multiplication of threats he could scarcely have imagined – it's time to salute those who unified a movement and created an organisation which is largely responsible for this success, and which remains as relevant as ever.

By the mid-1920s, the individual successes of the National Trust and Commons Preservation Society could not disguise that the majority of the countryside was facing a losing battle against a multitude of threats. 'Ordinary' countryside – blessed with neither commoners' rights, nor the beauty needed to make it an attractive acquisition for the National Trust – still had no national defender.

As early as 1908, a debate on 'The Disfigurement of the Country' had raged across the pages of *The Times* following a letter from Richardson Evans, an early benefactor of the National Trust and founder, in 1893, of the Society for Checking Abuses in Public Advertising (SCAPA). Evans responded to Sir Harry Hamilton Johnston's observation – after returning from thirty years exploring in Africa – that 'the beauty of the countryside is rapidly disappearing'. He urged 'all those who feel strongly about the degradation' of the countryside to 'act with each other' rather than 'vent their emotions in vain laments'.

Other correspondents noted that 'jerry-built' villas were being increasingly dotted around rural England, while 'rapidly growing outskirts' were creating 'wildernesses of brick and mortar'. The painter Delissa Joseph argued that the real evil was 'the permanent disfigurement of the beautiful English countryside by inartistic and unsuitable buildings', advocating the creation of a Ministry of Fine Arts for 'beautifying and preserving the countryside'.

JOINING FORCES

Richardson Evans believed combined action between the various, single-issue amenity bodies would create a network 'able to diffuse the knowledge of the better ideal'. In fact, this was a task Evans had first attempted in 1898, organising a conference for all those concerned with 'arresting the wanton destruction of picturesque effect or in preventing encroachments on the playgrounds of the people'. Evans' thinking was inspired by The Queen's Commemoration Open Space Committee, formed by Lord Hobhouse in 1896 to coordinate the work of existing preservation groups in dedicating over seventy open spaces around the country to Victoria's Diamond Jubilee. *The Spectator* felt that providing England 'with beautiful pieces of country which will be preserved for ever . . . in the nature of an oasis, however much the deserts of brick and stone may spread and increase' would be the most 'fitting memorial of the Victorian epoch'.

Hoping for similar success, Evans called on the House of Commons to set up a 'Parliamentary Group for Concerted Action in Defence of the Picturesesque and Romantic Elements in our National Life'. The intention was to 'form a non-party political force to strive for the wellbeing of others that will be more than a match for the philistines'. Evans succeeded in convincing his conference of peers, MPs and representatives of the preservationist movement to commit to 'asserting policies of maintaining beauty, simplicity, dignity and interest in the aspect of out-of-doors Britain'. However, a lack of widespread Parliamentary support, and the problem of forging a long-term union between fiercely independent campaigns conspired to scupper his plans.

PAGES 42–3
The kind of Gloucestershire valley that inspired the architect Guy Dawber to create a new organisation, to 'conserve what is beautiful in our countryside' and encourage 'the right type of development'.

IDEA 4 • CREATING CPRE

1914. Mr. William Smith answers the Call to preserve his Native Soil Inviolate.

1919. Mr. William Smith comes back again to see how well he has done it.

The post-war sprawl lampooned by *Punch* prompted one *Times* correspondent to highlight the absurdity of failing 'to raise a finger to thwart the designs of a speculative builder in a landscape made sacred by memories of Wordsworth'.

While Richardson Evans had failed to instigate a national movement to save the countryside in 1898 and 1908, conditions became more favourable for such an enterprise in 1918. The Great War had a huge impact on the way the English thought about their country and its future development. The trauma of war fostered a heightened appreciation of the finite nature of England's beauty, creating a wave of nostalgia for a rural golden age, and regret, in hindsight, for the damaging effects of industrialisation. After the war, the diminishing returns of local preservation societies and national 'single-issue' campaigns were starkly illustrated by a famous *Punch* cartoon showing a soldier returning to his village in 1919, only to find it had become a sprawling metropolis. A *Times* article of 1919, 'Rebuilding the Countryside: Plans in a Typical Rural County', captured the post-war zeal for development in describing Herefordshire county council as 'fully occupied with such topics as light railways, housing, good roads, drainage boards and the possibility of an extensive rural area being supplied with electric light and power'. In the same year, *The Times* warned that 'in our eagerness to create new wealth and provide new houses, there is a danger that we may sacrifice a rich inheritance of natural beauty', suggesting that a national Advisory Committee was needed to relieve the burden of the National Trust and 'exercise a strict surveillance of all that is worthy of preservation in England'.

In 1921, Sir Harry Hamilton Johnston once again sparked a national debate, calling for a 'crusade for the preservation of beauty in the English countryside'. By April 1922, *The Times* was speculating about the state of the countryside around London fifty years thence, 'when the present tendency towards universal suburbanisation will have come into full play. By then, how far will London have thrown out its tentacles?'

TURNING PROTEST INTO ACTION

The time had come for a coordinating force to unite the disparate arms of the preservation movement under a single banner. On 19 December 1925, Guy Dawber, President of the Royal Institute of British Architects (RIBA), wrote an impassioned plea to *The Times* for 'immediate steps to be taken to prevent the whole countryside being littered with architectural eyesores'. Dawber was deeply disturbed that 'on all sides we hear protests made, but nothing is being done to stop it. Every man is a law to himself and builds as

LAMBOURNE END
THROUGHOUT FARE 11ᴅ
From NORTH WOOLWICH

ROUTE 101
SUNDAYS
PASSING EAST HAM STATION, WANSTEAD, WOODFORD BRIDGE, GRANGE HILL, CHIGWELL ROW.

he wishes with a selfish disregard of his neighbour.' Initially proposing that 'the Ministry of Health [which had responsibility for housing] or some other body should call a conference to inquire into the whole matter', in January 1926 Dawber decided to take matters into his own hands. 'The time has come when definite steps should be taken to prevent the further destruction and disfigurement of Rural England,' wrote Dawber, to fifteen organisations with an obvious interest in the future of the countryside. 'The problem is a two-fold one: the conservation of what is beautiful and interesting in our countryside and towns and villages; and the encouragement of the right type of development.'

Dawber received a welcome endorsement from the Minister of Health, Neville Chamberlain: 'I sympathise very strongly with your endeavour to awaken public opinion of the importance of preserving rural amenities. I wish every success to the movement.' A month later, the President of the Town Planning Institute, Patrick Abercrombie, published his treatise on *The Preservation of Rural England*, encouraging Dawber and his supporters to form a 'National League for the Preservation of Rural England'. Abercrombie's aim was that this 'League' would 'focus attention upon the destruction of the beauty of the English Countryside which is not only robbing the country of one of its most precious assets but is uneconomic'.

Dawber and Abercrombie put this plan into action with a series of meetings throughout 1926. Lord Crawford, a sponsor of the Ancient Monuments Act of 1913 and Chairman of the Fine Arts Commission,

The rural Essex landscape captured in Dorothy Paton's poster of 1926 — the year of CPRE's formation — has changed remarkably little in the past ninety years, thanks in no small part to the organisation's campaigning. Fittingly, the village's outdoor learning centre was featured in a 2010 CPRE report on the positive impact of Green Belts.

CPRE's founding logo designed by Frederick Landseer Griggs, an executive member of the National Trust and one of the first etchers to become a full member of the Royal Academy.

was co-opted as President of the new body; having recently turned down a ministerial post in Stanley Baldwin's Government, Crawford would soon decline the chance to become Chair of the BBC, such was his concern for the countryside. A draft constitution was drawn up, giving what was now a Council rather than a League, a specific remit to 'secure the protection of rural scenery' while acting as an advisory body 'to arouse, form and educate public opinion'.

In August, *The Times* declared:

> The campaign for the preservation of the beauties of rural England is to be further assisted by the establishment of a central body which will be representative of the numerous societies at present at work, and form a propaganda organisation to coordinate the activities of all. The hope is that by drawing them all together in one representative council, it may enable them to take quicker and more powerful action. England must grow and houses must increase. But it need not grow just anyhow, and the new houses need not be dumped down just anywhere.

THE BIRTH OF THE COUNCIL FOR THE PRESERVATION OF RURAL ENGLAND

Guy Dawber introduced the first public meeting of CPRE on 7 December 1926, 'inaugurating a movement which, we hope, will result in the preservation of rural England, the saving of that national treasure of beauty which means so much to all of us, and which we see threatened with imminent destruction'. Dawber felt the new Council – which included the National Trust, Women's Institutes, Commons Preservation Society and SCAPA – represented a coalition the like of which had 'never before been brought together for a great common purpose'. Neville Chamberlain then addressed the meeting to welcome the formation of 'a body of authoritative character', which could 'draw attention to threats against specific beauty spots' and 'offer local authorities technical advice and assistance'.

Press reaction to the new Council's formation was wholly positive. *The Times* felt that 'the existence of so strong a combination can hardly fail to impose a check on the spirit of careless vandalism which has already done so much harm throughout the country'. *The Spectator* ventured that 'the Council underrates rather than exaggerates the threat against the most characteristic charms of our landscape', but found a good omen in the number of its founding constituent bodies: 'twenty-two – the length of a chain or cricket pitch, the unit of the square acre – is quite the most English of all the numbers.' *Country Life* said that CPRE was the public expression of the anxiety felt by all educated people for the future of the countryside. *Punch* noted that it was widely accepted that CPRE had begun well and were 'not letting the grass grow under their feet', before pondering 'but we thought this was what they wanted to do!' By February 1927, the *Manchester Guardian* was reporting that the Council was 'being overwhelmed by letters of protest directed against the erection of new buildings and the hundred and one structures which follow the extension of popular motoring'.

WHAT WAS SO DIFFERENT ABOUT 1926?

The year 1926 saw the General Strike, the first transatlantic telephone conversation and the first demonstration of television by John Logie Baird. And yet the formation of CPRE was ranked as one of the top ten news stories of the year by the *Guardian*'s 'Century' archive project; Abercrombie's manifesto was ranked as one of the five key publications of the year by the British Library's 'Chronology of Modern Britain'. The formation of an 'umbrella group' for the preservation movement in such a tumultuous year for Britain might have been expected to encourage allegations of insulting arrogance or, at best, insensitive naivety. Instead, the speed of CPRE's formation, and the universal acclaim and goodwill it received, suggests its founders were in possession of rare vision – a vision that managed to unite a country which had been threatening to go to war with itself.

Until 1926, there had been an inherent problem with the idea of preservation: the lack of a programme for change that would not just attempt to take us back to the rural utopia of William Morris' *News from Nowhere*. The founders of CPRE were a new breed of preservationists, who believed we could enjoy the benefits of industrialisation while curbing its most damaging side effects; development did not have to be destructive, and preservation did not have to be a barrier to progress. With Dawber, Abercrombie and Crawford, the Romantic's appreciation of nature and heritage combined with the Modernist's zeal for advancement. The historian Sir Max Hastings (CPRE President, 2002–7) has suggested that society reached a tipping point in 1926. Enough 'people recognised the unacceptable aesthetic price that was being paid' to maintain the orthodoxy of unfettered industrialisation, and wished to find an alternative to what Sir Kenneth Clark called 'heroic materialism'.

On the eve of CPRE's formation, *The Times* noted the admirable work of the National Trust, but conceded that 'it does not cover more than a fraction of the ground' and could lead to 'patchwork protection – the saving of a handful of places of exceptional natural beauty' at the expense of the rest of the countryside. The implication was that a body like CPRE was imperative if the importance of '*every* wood, *every* stream and *every* hedgerow' was to be recognised and defended.

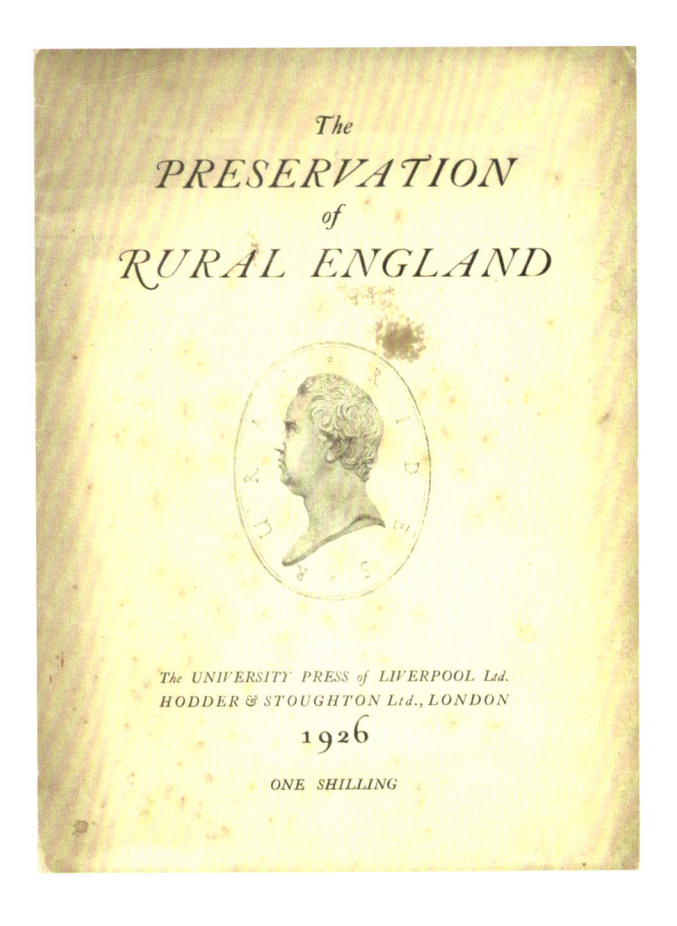

ABOVE
Patrick Abercrombie's 1926 vision for the future of the countryside led to the formation of CPRE within months of its publication. The cover featured a tribute to William Cobbett's *Rural Rides* of a century earlier.

OPPOSITE ABOVE
A CPRE poster from the mid-1960s showed how it had evolved from being a coordinating force, to having a distinct identity as the guardian of England's countryside.

OPPOSITE BELOW
When CPRE celebrated its Jubilee in 1976 the artist John Piper donated this special design to promote the 'Save Your Countryside' appeal.

By 1928, the journalist and tennis commentator J.C. Squire could confirm that Richardson Evans had been right all along:

> A multitude of Societies have come into being, ranging from the Society for the Preservation of Ancient Buildings to the Pure Rivers Society. CPRE is a rival of none of these bodies: it at once federates and transcends them, doing, in some measure, the work of each of them without obliterating their individual functions, whilst, by the virtue of its size and scope, strengthening considerably the forces which at any particular moment have to be brought into action on any particular threatened front.

The following year, the three party leaders took time out from the 1929 General Election campaigning to endorse a fundraising appeal 'to assist this worthy movement'. Baldwin, MacDonald and Lloyd George united to advocate CPRE's work to preserve the countryside, in the hope that it could 'effect necessary changes to avoid injuring a precious heritage'. CPRE had arrived, and the preservation movement was at last able to show its real strength by placing its individual concerns within the context of the future of the whole countryside. Meanwhile, the ideas contained in Abercrombie's manifesto would ensure that CPRE would soon become much more than a coordinating committee, but a campaigning force with its own identity and aims.

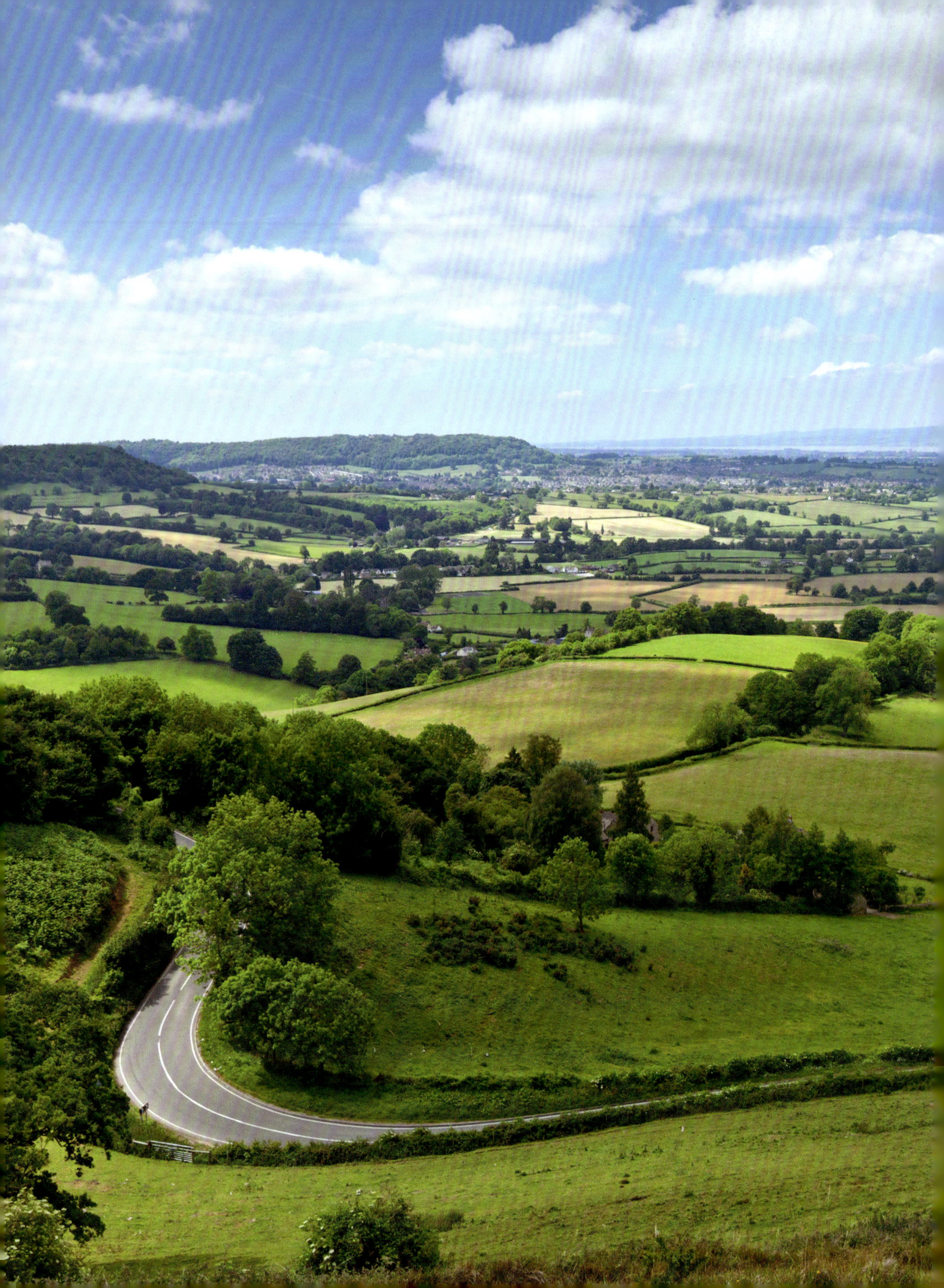

Idea 5

Taming the Octopus

introduced by Jonathan Dimbleby

Hardly anyone now uses the phrase 'ribbon development'. In large measure this is thanks to a campaign first mounted by CPRE over eighty years ago. It was a battle to prevent every road linking one town to another being enclosed on each side by a 'ribbon' of houses, shops and factories. The campaign highlighted that this growing blight threatened to deny future generations the chance to relish the beauty of the countryside from inside the car as we do today.

The campaign not only helped to arrest 'ribbon development' directly through an Act of Parliament in 1935 but also pioneered the values and principles which underpinned subsequent Town and Country Planning laws and the creation of Green Belts as national policy sixty years ago.

Nothing could better illustrate CPRE's potential to protect the present and to enhance the future character of our precious landscape. Never has the role of CPRE been more crucial.

One of the catalysts for the formation of CPRE had been the phenomenon of 'ribbon development' – a new form of urban sprawl, which radiated out from urban areas along 'arterial' roads. It is hard to imagine that a term that has faded so deeply into obscurity was a national scandal in the early 1930s, dominating debate in Westminster and Fleet Street. Back in April 1925, a year before the publication of *The Preservation of Rural England*, Patrick Abercrombie urged Neville Chamberlain, the Minister responsible for housing, to regulate 'residential growth [by] the ribbon unrolled along the roadside . . . the new method which is unconsciously being adopted through the whole of England as a result of the new motor-omnibus services and use of private motor-cars.'

Abercrombie lamented that 'soon this green and pleasant land will only be glimpsed through an almost continuous hedge of bungalows and houses!' Not only would ribbon development 'inflict maximum destruction on rural beauty', but it also represented 'the most extravagant method [of housing] to sewer, water, light and police'. By the end of 1926, Chamberlain's speech at CPRE's inaugural meeting would acknowledge that the 'spoiling of undefiled countryside by ribbon development' was the biggest 'aesthetic abomination' facing rural England; 'besides being undignified, if not positively offensive, it is also uneconomical, wasteful and inconvenient.' The war correspondent Sir William Beach Thomas wrote that it 'had become an urgent practical necessity that the coarse ribbons should not extend in indefinite disarray along the roads', while CPRE promised 'to attack it from every possible angle'.

THE DEBATE BEGINS

The Parliamentary angle had been ineffective until, in February 1929, a passionate speech from the young Middlesbrough MP 'Red' Ellen Wilkinson started six years of public debate on the issue. The former communist – who was still reported to sleep with a portrait of Lenin over her bed – argued that legistlation must be used to bring 'backward' authorities into line:

> Anyone who takes a motor drive along some of these arterial roads will see the perfectly miserable conditions which are growing up, with an utter lack of beauty and design. If you drive along the main roads in Germany you will see that they have taken the houses away from the main roads and grouped them in beautiful small villages, which are linked up by smaller roads with the main thoroughfare. Why could not that be done in England?

Wilkinson was ably backed by personal evidence from another Labour MP, Richard Wallhead: 'I have not lived in Hertfordshire long. When I went there first it was a pleasant ride from my home to St Albans. Now it has become a nightmare and an atrocity. The whole place is built up from end to end with nothing but a string of unaesthetic little bungalows,

PAGES 50–1
Saved from ribbon development, this Cotswolds landscape is just one of many in rural England to have benefitted from the campaign to tame the 'octopus'.

The Cornish countryside largely escaped the ribbon development that plagued much of England in the 1920s and 1930s.

which are a positive eyesore.' By 1932, the future 'austerity' Chancellor, Sir Stafford Cripps, hoped the new Town and Country Planning Bill would address the ribbon development which he felt had 'led not only to a tremendous depreciation of the beauty of the countryside but also to traffic difficulties and dangers'.

The eventual Act failed to deliver, and by the summer of 1934 the problem was becoming intolerable; CPRE's Lancashire and Wiltshire branches joined the Oxford and Cambridge Preservation Societies in publishing reports on the matter. In July, CPRE organised a conference at the Middlesex Guildhall in Westminster, at which county surveyors, builders, landowners and motoring organisations all agreed that ribbon development needed dealing with through new legislation. As the pages of *The Times* were flooded by complaints about ribbon development, one correspondent quoted Cowper's poem of 1782, *Retirement,* to show that its 'disadvantages were recognised even 150 years ago!'

> Suburban villas, highway-side retreats,
> That dread th'encroachment of our growing streets,
> Tight boxes, neatly sashed, and in a blaze
> With all a July sun's collected rays,
> Delight the citizen, who, gasping there,
> Breathes clouds of dust, and calls it country air.

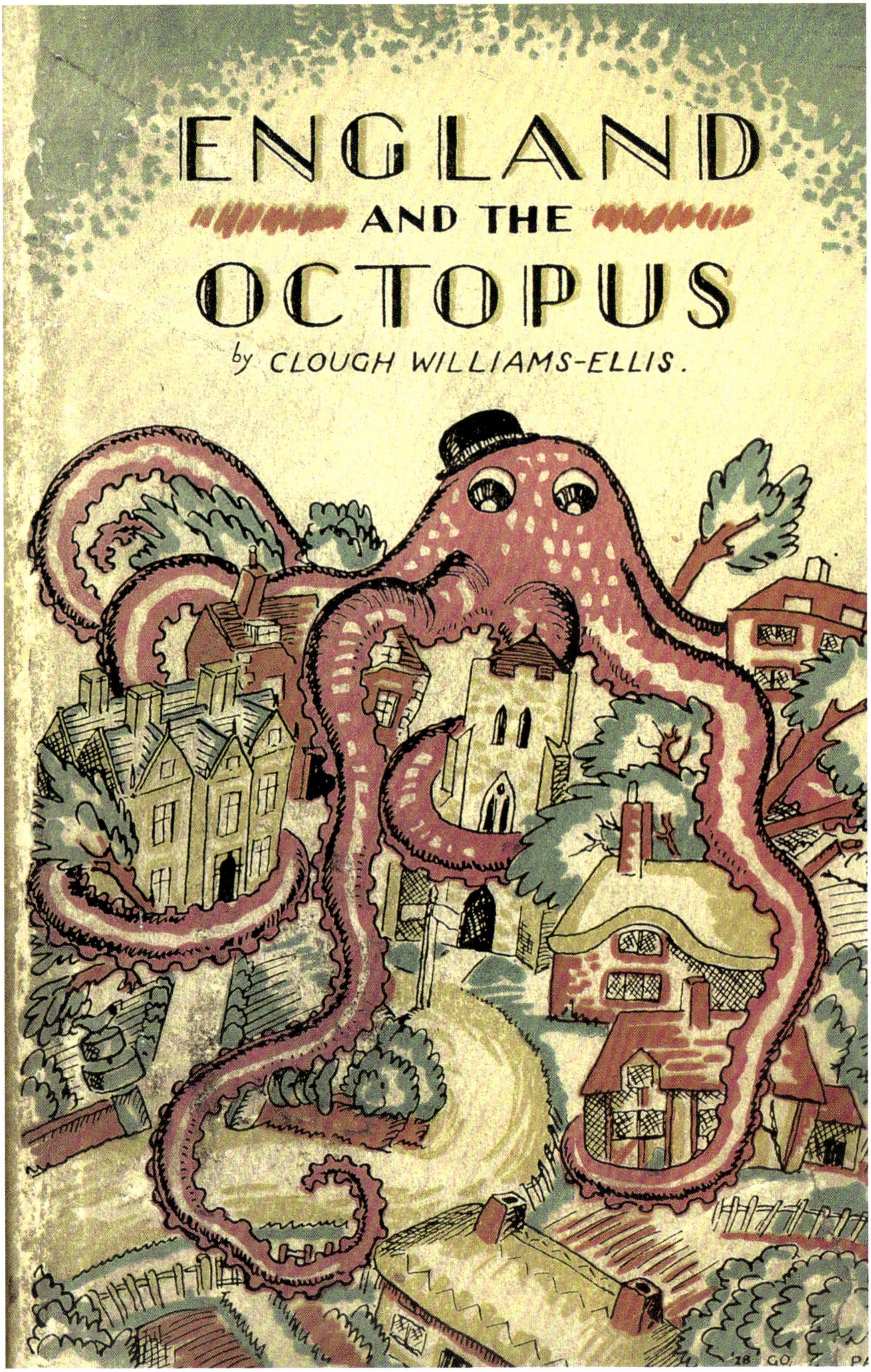
ENGLAND
AND THE
OCTOPUS
by CLOUGH WILLIAMS-ELLIS.

OPPOSITE

England and the Octopus helped make ribbon development a high-profile issue from 1928, and was an influence on a generation of writers including D.H. Lawrence, Evelyn Waugh and George Orwell. It has been described by the philosopher Roger Scruton as 'one of the most fruitful of all the pre-war attempts to conserve the many managed environments of England'.

The following day, a *Times* leader gave a visceral endorsement of CPRE's calls for fresh legislation:

> Ribbon development turns enormous stretches of roads into death traps. It scars the naturally lovely face of the countryside with great slashes of ugliness. These things have been tolerated far too long; and the remedy is not far to seek. But why should the matter be left to sporadic legislation? The case is clear enough for the Government to introduce a Bill.

A NATIONAL EMERGENCY

In October 1934, the Ministers of Health and Transport (Hilton-Young and Hore-Belisha) issued a joint public statement to 'welcome, and desire to encourage, the activities of CPRE in stimulating a sound public opinion upon the question of Ribbon Development'. CPRE's President Lord Crawford needed no further invitation to write to Prime Minister Ramsay MacDonald demanding that 'emergency legislation be passed' to prevent ribbon development turning 'country roads into streets . . . changing the social structure and raising the cost of living. The future shape of the country is at stake.'

Crawford's fear of roads turning into streets was influenced by the most unlikely source: *Lady Chatterley's Lover*. As Connie was driven through the Nottinghamshire countryside, D.H. Lawrence described how the new 'miners' cottages . . . stood flush on the pavement . . . lined all the way. The road had become a street, and as you sank, you forgot instantly the open rolling country.' Lawrence had, in turn, been inspired by the Portmeirion architect and chief propagandist for CPRE, Clough Williams-Ellis. He wrote a glowing review for *Vogue* of Williams-Ellis' 1928 polemic *England and the Octopus*, a book commissioned by CPRE as a provocative declaration of war on the ugliness insidiously creeping across the countryside.

The cover illustration famously showed a bowler-hatted octopus (representing London) stretching its tentacles all the way into a rural village – a classic case of the ribbon development described by Williams-Ellis as 'disfiguring little buildings' which 'grow up and multiply like nettles along a drain, like lice upon a tape-worm'. Williams-Ellis' use of the octopus imagery was a masterstroke – suddenly, ribbon development felt even more sinister – and the book was hugely influential. Evelyn Waugh's *Vile Bodies* of 1930 had characters feeling physically sick in the face of ribbon development: 'a horizon of straggling red suburb; arterial roads dotted with little cars; factories, some of them working, others empty and decaying'.

Crawford's letter to Ramsay MacDonald had the desired effect; the King's Speech of 1934 stated that 'measures to control ribbon development would be introduced if time permits'. Stanley Baldwin responded to this prevarication by telling MacDonald: 'There is one very important Bill . . . for dealing with what, I think, is a most urgent and difficult matter –

the problem of ribbon development. I think it is essential that the House should get down to that question.'

With the threat of new legislation encouraging developers to make hay while the sun still shone, the question of when legislation would be introduced dominated Prime Minister's Questions in early 1935. Viscountess Nancy Astor joined in the harrying, asking 'the right honourable Gentleman [to] go down the Great West Road where he will realise that, if he does not hurry, [legislation] will be of no use?' Within a fortnight, MacDonald announced that a Restriction of Ribbon Development Bill was in preparation, to the relief of the Bishop of Winchester, who cited the 'really detestable ribbon development of horrid little houses and wretched advertisements' which had blighted the views of Oxford's dreaming spires.

When the Restriction of Ribbon Development Act was finally passed in July 1935, highway authorities were allowed to preserve views and amenities by controlling building within 220 feet of roadsides. Most importantly, the success fully justified the formation of CPRE by showing it had the ability to build widespread public and political support for its aims. Of course, ribbon development continued until local authorities got to grips with their new powers, but it is a measure of the success of the Act that it was a largely forgotten problem by the onset of the Second World War. By 1976, the then *Country Life* editor Gordon Winter celebrated CPRE's golden jubilee by noting that it was 'their clamour and agitation' that meant ribbon development was a 'horror of the 1920s and 1930s that we have largely forgotten'.

ABOVE
This tea shack in Buckinghamshire was typical of the structures springing up along roadsides in the early 1930s.

OPPOSITE
The Act of 1935 finally gave councils the powers to deal with the problems of speculative ribbon development, allowing them to acquire land up to 220 yards from a highway to prevent 'the erection of buildings detrimental to the view from the road'.

THE ROADSIDE

WHY NOT BUILD YOUR OWN HOUSE HERE?

The answer to this question, which is reiterated all along the St. Albans Road, (quite apart from the amenities of any house on such a site,) is that building along arterial roads impedes the flow of traffic, that in the long run it is the most wasteful kind of development, and that it cuts the road passenger entirely off from the countryside.

Another typical scene on the Colchester Road. The fringe of the road between London and Chelmsford is almost entirely built over or for sale.

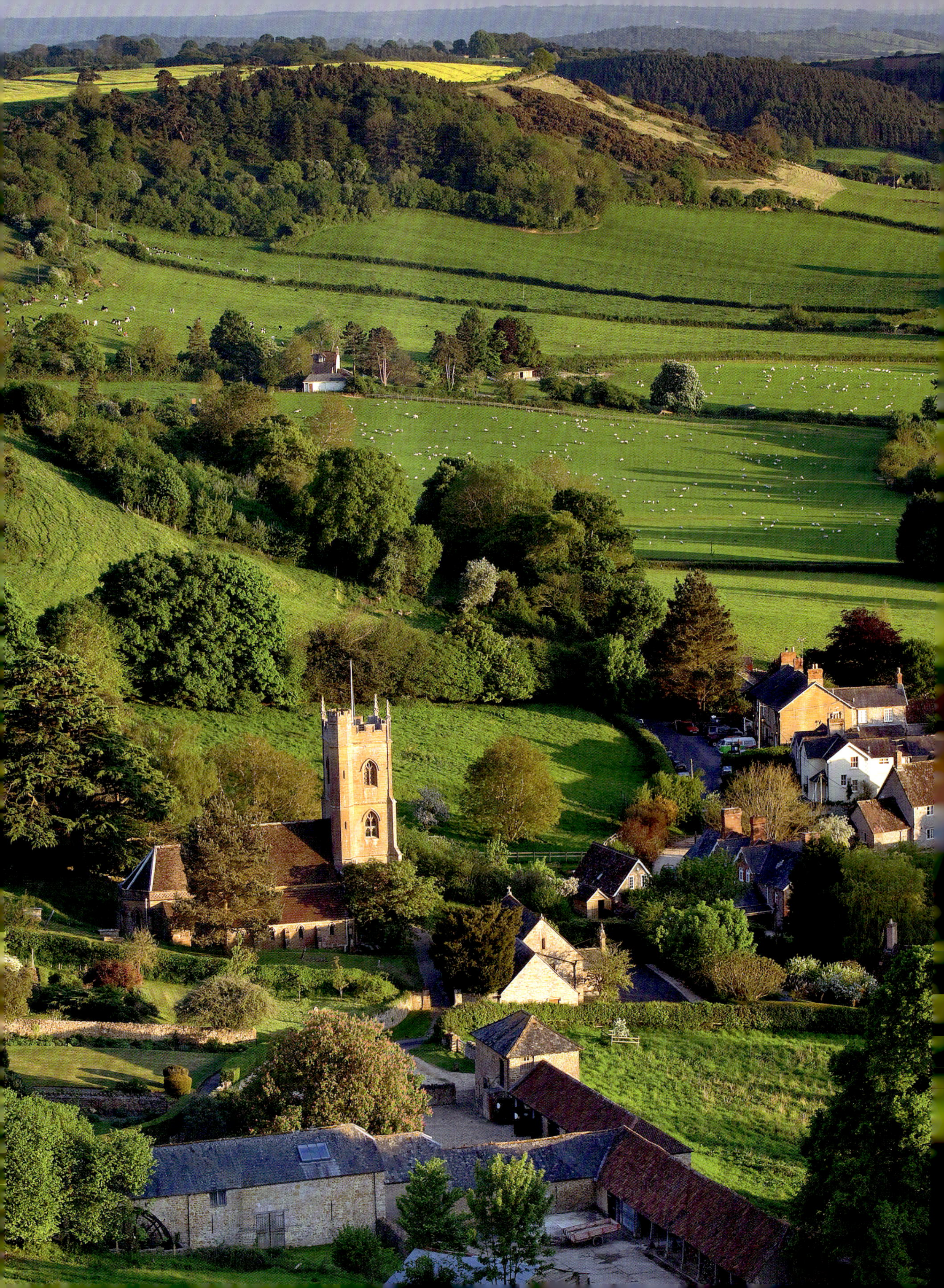

Idea 6

Keeping Villages Alive

introduced by Kate Adie

We love our countryside. Artists celebrate the landscape, visitors rejoice in weekend visits. But that's not enough. We have to live there, make it come alive. And the village is the heart of that life. Villages reflect our history: Bronze Age settlements perched defensively on hilltops, medieval hamlets serving the lord of the manor, busy clusterings by rivers for millers and traders, pitmen's Victorian terraces on the coal seams. Villages today need good transport links and the Internet, sympathetic planning and intelligent building design and energy use. There's no future in being a retirement home with sheep. The pub, the shop, the church, the school: all are under pressure. But, young and old, families and farmers, do not snooze among the hedgerows, but accept that you must adapt to survive. Long live the village!

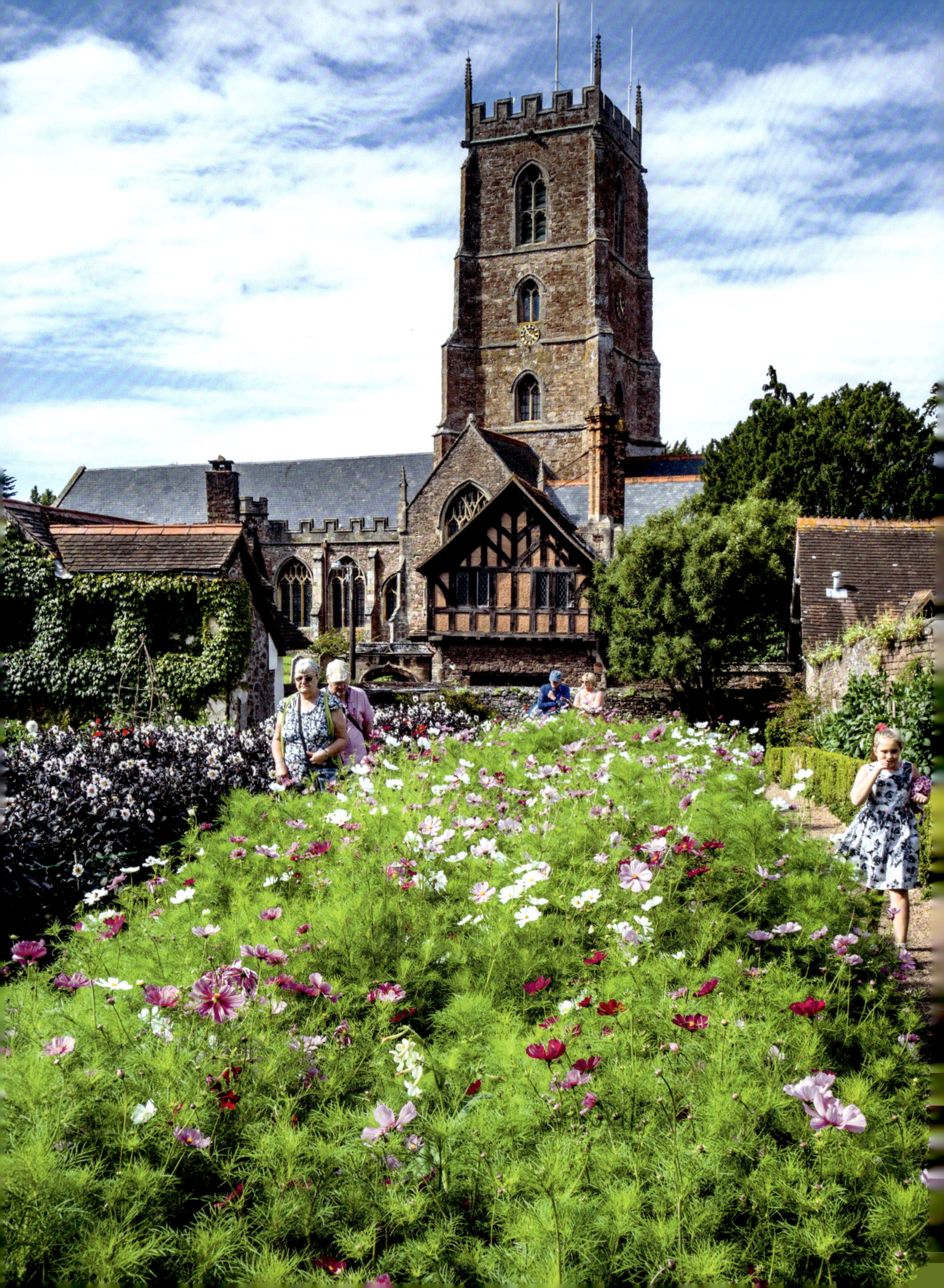

PAGES 58–9
The picturesque village of Corton Denham in Somerset is living proof of the success of the campaign to save England's villages.

OPPOSITE
Dunster in Somerset may have been the kind of place *The Spectator* had in mind when it appealed to the National Trust to secure the 'ideal village' (with 'cottage gardens ablaze with colour') for posterity, such was its concern for the speed at which declining villages were being sold off by 'impecunious squires'.

Although England's transition to a predominantly urban population was complete by the 1851 census, the acceleration of the process created a rural exodus that threatened one of the defining aspects of the countryside with extinction. The English village had always been the lifeblood of the landscape – a hub of activity providing workers for the land and stimulating the commerce which kept rural farms and families alive, and allowed market towns to thrive. But a late-nineteenth century combination of mechanisation and urbanisation meant that many of the rural jobs that supported villages moved to the towns. Meanwhile, food imports from the North American prairies combined with the dominant free trade ideologies to create a long-term agricultural depression. As the historian Victor Bonham-Carter put it, 'home farming was deliberately abandoned to sink or swim.'

Incredibly, the amount of England used for arable farming fell by around a quarter between 1879 and 1900, accompanied by a 50 per cent drop in farm rents. By 1902, around one in five farms was left unoccupied after cheap food imports exacerbated a series of poor harvests, with the rest in an increasingly decrepit state. Doubts were even raised about the very future of the village as a viable form of community; Bonham-Carter argued that 'by 1914, at the latest, the heart had gone out of the village as an economic unit'. Without thriving villages, England faced a bleak future of neglected landscapes and a dependency on food imports.

A NEW SENSE OF COMMUNITY

Grace Hadow had been a campaigner for women's suffrage and helped create England's first Rural Community Council in Oxfordshire in 1920, promoting pioneering adult education schemes through village libraries or 'reading rooms'. Between 1918 and 1921, 400 Village Clubs were also formed by Sir Henry Rew, who feared that the war had heightened class differences, and aimed to bring all villagers together socially. He hoped his clubs would awaken 'the social consciousness, a development of the true democratic spirit', venturing: 'I cannot imagine a gentleman who has bowled out the village blacksmith not having a fellowship towards him afterwards.'

Of course, the social activities organised by the Rural Community Councils and Village Clubs Association (VCA) needed an alternative venue to the village pub, vicarage or tithe barn. Village halls were a new and desperately needed solution. One farmer's wife approached the VCA to help her village secure one out of concern for her sons, newly returned from the trenches: 'Now that they are back I have far more anxiety about them than I had all the time they were fighting. In the evenings they have nowhere to go except the roadside or the public house, and they are rapidly going to pieces.' In 1932, the National Council of Social Service noted that 'a good village hall is the key to vigorous social life'. Many villages converted old barns to halls, while others made do with cowsheds

'Unless life in villages is worth living, the younger and more enterprising villagers will not remain in them.'

furnished only with dry straw. Many more built new village halls with financial help from the VCA and Rural Community Councils.

In 1922, Prime Minister Lloyd George warned that 'unless life in villages is worth living, the younger and more enterprising villagers will not remain in them', prompting *The Times* to ask 'what can be done to stop the drain of the rural populace to the towns?' A 1923 Mansion House meeting saw Lord Lincolnshire acknowledge that the need 'to promote the wellbeing and contentment of village life by reviving the old community spirit in the countryside' was of 'urgent importance in the national interest'.

While a 1924 *Times* leader on the 'Revival of Village Life' lamented 'rural depopulation' and noted that folk dancing could help relieve monotony, there was no guarantee that dancing would achieve Sir Henry Rew's aim of breaking down social barriers. VCA minutes reveal that one village hall dance was a failure because 'the wife of the upper-butler would not dance with the under-butler; the wife of the upper-gardener would not dance with the under-gardener, and as for the domestic servants – they are a class apart!'

Regardless of the inevitable failures of decorum, there was a growing demand for dancing and theatre in rural communities. By 1926, around 3,000 villages had written to the Village Drama Society for advice on using their halls for amateur dramatics. Mary Kelly had founded the society in her Devon village in 1919; by the late 1930s, 600 villages around England were members of the society, while her 1939 book, *Village Theatre*, became influential in promoting village drama as a way to inspire villagers to celebrate their unique local customs and traditions.

REJUVENATING RURAL HOUSING

As well as being keen supporters of village theatre, the Women's Institutes also campaigned keenly for vital rural services, including housing. In fact, in 1918, they passed their first ever resolution to encourage local authorities to build new homes in villages, with support from Sir William Savage, who went as far as calling the decrepit state of most rural housing 'a disgrace to the country'. Despite the interwar housing boom, very few rural workers benefited from new housing, with nine out of ten remaining in poor quality tied accommodation in 1939.

Sleepy scenes of village life from the 1920s, when ending the migration of the rural population to cities became an issue of national importance.

IDEA 6 • KEEPING VILLAGES ALIVE

Better housing was the main focus of CPRE's wartime campaign to improve 'social amenities for the rural population' at a time when only 7 per cent of farms had electricity and a third of England's parishes had no piped water. The wartime revival of farming further increased the demand for rural housing which was already being driven by the tendency, noted in CPRE's 'Rural Housing' pamphlet of 1932, for wealthy urbanites to 'snap up' desirable cottages for weekend use. In the pamphlet's foreword, the leading planner and architect Raymond Unwin looked forward to a time when 'the rural worker will be able to meet his urban brother on fair terms in securing houses', urging 'the protection of the rural workers' nests from the cuckoo-like invasion of the weekender'.

By 1945, the campaigning of the WI and CPRE encouraged the Attlee Government to increase subsidies for the rural housing required to house the 100,000 permanent workers now needed on the land; Health Minister Nye Bevan told rural councils that 'unless they build houses they will lose their workers'. By May 1946, 10,000 new rural homes

War memorials became a focal point for villages like Guiting Power in Gloucestershire after the Great War. After 1945, CPRE Gloucestershire's President, Lord Bledisloe, proposed that 'no form of war memorial will appeal more to our fighting men when they come home than something in the nature of a Community Centre or a village hall'.

Village playing fields became highly sought after in the Second World War, but boasting England's largest village green (at 43 acres) meant Great Bentley in Essex could continue to host cricket matches like the one captured here by Walter Bayes in 1940.

were under construction, and during the winter of 1948–9 *The Times* ran a series of illustrated features on the new schemes springing up in villages around the country. In June 1949, their leader on 'Rural Realities' reported that 'a new spirit now happily prevails, and village halls, houses, drainage and electricity are being treated as natural requirements'.

FOSTERING VILLAGE PRIDE

Amidst the housing boom, care needed to be taken to ensure that community spirit was not damaged by the influx. In the interest of patriotism and community, CPRE made the Coronation of Queen Elizabeth II in June 1953 the hook for new campaigns for commemorative village greens and village hall extensions, perhaps featuring 'a billiards room, or a club room suitable for the young people of the village'. A 'Tidy Village Competition' was the third element of CPRE's Coronation celebrations and by 1957 their branches' new 'Best Kept Village' competitions were being held in most counties, doing 'much to interest the residents in their surroundings and to encourage them to take pride in them'.

War memorials had been an earlier focus of village pride after, in January 1919, the architect Herbert Baker suggested that well-designed memorials to the victims of the Great War could 'become the centre of some beautiful, restful place . . . to express the heritage of unbroken history and beauty of England which the sacrifices of our soldiers have kept inviolate'. Then, in 1937, another wave of national memorials arrived in villages through the scheme to create playing fields in honour of King George V. During the Second World War, the National Playing Fields Association protested that forty-five King George's Fields were being used for training troops, while another forty-four had been ploughed up for growing food. CPRE was adamant that the fields should be maintained for recreational purposes; the Bolton MP Major Sir Edward Cadogan agreed, pointing out that, in commandeering playing fields, we were in danger of creating a 'brown and unpleasant' land on the very 'playing fields upon which, who knows, we may win further Waterloos'.

War memorials, together with playing fields, village halls and decent new housing, helped save the English village as a viable form of community. If not for the efforts of the campaigners who fought to ensure they remained the social hubs of the countryside, it is hard to imagine that villages would have survived in anything like the form we know today.

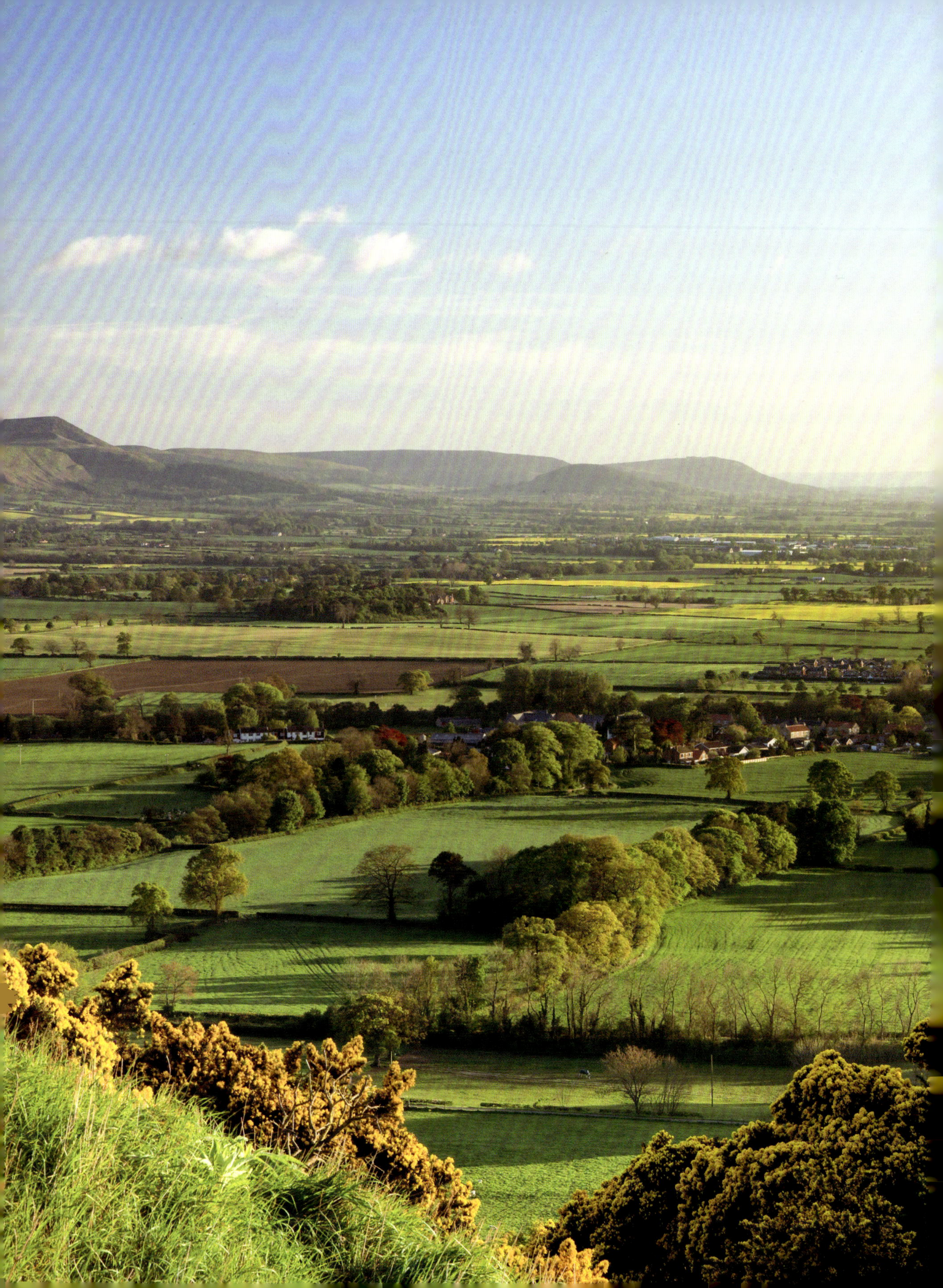

Idea 7

Rural Planning

introduced by Fiona Reynolds

Fly over the east or west coast of America, the Mediterranean or Turkish coasts, and the thing that strikes you above all is . . . sprawl. We risk taking for granted the protected coastlines and clear separation between town and country that makes the English countryside so beautiful and highly valued today.

Yet that fate was so nearly ours. Were it not for the energetic defence of England's beauty, and determined campaigning, the tide of development would almost certainly have swept throughout southern England and from the cities of the north. Patrick Abercrombie, the architect of rural planning, warned that 'this rural England of ours is at this moment menaced with a more sudden and thorough change than ever before,' and that 'it is not safe to leave these changes to adjust themselves, hoping that somehow a general harmony will result from individualistic satisfactions'.

The achievements of the 1920s campaigners were planning principles and practices that made us the envy of the world. And so today, when planning is castigated for holding back development, let us remember and honour their work – it has saved England's beauty.

Until 1926, 'planning' meant town planning; simply, the idea that settlements would be built up with some thought to their location and layout. Evidence of town planning can be found in the earliest known cities of ancient Egypt and Mesopotamia, and one might even say that it was the first defining quality of civilisation. And yet, until the early twentieth century, virtually no consideration whatsoever was given to planning or controlling what was built outside cities. Of course, there was little need to. Nothing much was built outside cities; the occasional rural settlement, farm building or place of worship, certainly nothing to warrant any kind of regulation by a ruling power. The ancient Britons, Saxons, Romans, Normans and the English aristocracy have all left their mark on the countryside, albeit in a generally complementary way. If steam-age railway infrastructure is now seen as the last golden age of large-scale human interaction with the landscape, it seems all the more attractive because of the unplanned sprawl that came after it.

PEACEHAVEN AND THE PLOTLANDS SCANDAL

Patrick Abercrombie's 1926 inventory of threats to the English countryside in *The Preservation of Rural England* helps us understand why he had used the same treatise to announce his invention of 'rural planning' (even subtitling it *The Control of Development by Means of Rural Planning*). The specific problems identified by Abercrombie were the growing tendency of industry to occupy rural areas (through the building of new factories, quarries and mines); the rise of 'ribbon development' along country roads; and the trend for 'weekend colonies' of huts and bungalows to appear in the middle of the countryside. The astonishing genesis of Peacehaven was a symptom of the latter, also known as the 'plotlands' phenomenon.

Plotlands were pieces of land in attractive locations, divided and sold by speculators to individuals who wanted a bungalow in the country. An inevitable response to falling farmland prices and the desire of the aspirational working classes to escape the city, they provided a lifeline for farmers, a glorious opportunity for people craving countryside, and the chance for enterprising land agents to make a killing. Unfortunately, the unviable farmland that was sold off was usually in coastal locations – some of the most attractive and unspoilt parts of the country.

In 1916, the land agent Charles Neville used a *Daily Express* competition to promote plots he had acquired on the Sussex coast as 'free' prizes; but when he tried to charge winners a substantial 'conveyancing fee' the paper sued. And it was not just the paper who sued; some of those who had paid their £3 3*s* fee found that the promised beaches were 100-foot cliffs, while sewage from Brighton flowed into the sea near by. Nevertheless, the free publicity for what was then known as New Anzac-on-Sea was priceless, and by 1924 there were 3,000 people living in England's version of the Wild West. The Sussex Downs' own 'frontier'

PAGES 66–7
This view of the Cleveland Hills in the North York Moors National Park shows why Patrick Abercrombie believed that skilful rural planning could allow the countryside to absorb development without losing its beauty.

IDEA 7 • RURAL PLANNING

Peacehaven's proprietor Charles Neville was a master of marketing, but his promotional postcards showing a bucolic seaside resort didn't quite reflect the reality of what was being described as 'a monstrous blot'.

town – renamed Peacehaven after another competition in 1917 – had no roads, power or sanitation for many years, and was described by the planner Thomas Sharp as 'a national laughing stock'. It was no laughing matter for the broadcaster Howard Marshall, who thought it a 'monstrous blot on the national conscience'.

Peacehaven was the catalyst for Abercrombie's idea of Rural Planning. As he put it on the second page of his manifesto, 'the very people who are colonising the country in order to escape from town conditions are themselves helping to destroy the countryside through lack of foresight.' Two years later, in *England and the Octopus*, Clough Williams-Ellis raged about the worst kind of 'plotlanders': 'they care no more for the countryside they are billeted on than barbarian invaders are wont to do. Peacehaven may be cited as the classic example of the ravages of this distressing and almost universal development.'

Because they were largely self-built, places like Peacehaven were initially dominated by caravans, shacks and even bungalows sold in kit form; those that wanted more room even acquired old railway carriages to put on their plots. Abercrombie refers to 'wooden shanties – on wheels in order to avoid rates', whose 'temporariness' would quickly reduce them to an ever-greater state of dilapidation. Or, even worse, the opposite would happen; the weekenders would make the permanent move and demand proper roads, telegraph poles, municipal tips and all the rest.

POSITIVE PLANNING CHALLENGES PRIVATE RIGHTS

Abercrombie argued that with 'skilful planning' the countryside could absorb 'more buildings, new roads . . . and yet preserve its beauty'. A *Manchester Guardian* article of September 1926 lamented that the fledgling CPRE had not been formed in time to get Peacehaven's 'layout placed in the hands of the most enlightened experts on town-planning and architecture'. The article gave a sense of the scale of the task facing Abercrombie and CPRE, conceding that 'public opinion is not yet sufficiently advanced for such an interference with private rights' as would be needed to 'prevent anyone putting up a house utterly out of harmony with its environment'.

Abercrombie's idea of rural planning was intended to create a harmony between town and country by extending across rural areas, the protection given to 'public amenity' – defined as areas of natural beauty and historic interest – by the Housing and Town Planning Act 1909. Abercrombie argued 'it is high time to drop the clumsy word "town-planning" as applied to the country. The term therefore proposed is Rural Planning.' In late November 1926, two weeks before CPRE's inauguration, their committee member and Liverpool MP Sir Leslie Scott outlined the case for legislation, arguing that 'no man is entitled to use his land to offend the public by destroying the beauty of its surroundings. An aesthetic nuisance ought to be subject to legal control just as much as any other.'

The Times supported Sir Leslie, attacking the 'stupidity and anarchy which allows a dozen bungalows to destroy the natural features of a square mile of beautiful country', and anticipating Lord Goodman's words prefacing *The Sack of Bath* in 1973: 'stupidity is even more difficult to control than evil.' Two days later a remarkably prescient letter from Constance Giffard asked whether there could not be 'some authority from whom permission must be obtained before any building is begun, and through whom the proposition should be advertised publicly so that objections could be raised before it is too late?' Prompted by her regular drive through the Shropshire Hills being spoiled by new houses appearing 'with alarming rapidity, like a mushroom, almost in a night', Ms Giffard instinctively proposed what Abercrombie hoped would become basic principles of Rural Planning.

REBUILDING SHAKY FOUNDATIONS

The Town Planning Act of 1909 had encouraged lower density housing to improve sanitation, allowing towns to sprawl into what Abercrombie felt were 'monotonous' suburbs of 'interminable semi-detached villas . . . lacking architectural grace and interest of layout'.

The author of *The Rise and Fall of the Victorian City*, Tristram Hunt, has argued that the Act 'signalled the shift from urban to suburban living', resulting in 'an avalanche of concrete across the countryside'. Abercrombie's response was to insist that 'the country must not be regarded as a mere sleeping partner in this process of change and spreading of urban conditions'.

The path to more countryside-friendly legislation required the kind of political support which might have been expected from the Conservative Prime Minister Stanley Baldwin, who had recently asked 'what England may stand for in the minds of generations to come, if our country goes on during the next generation as she has done in the last two in seeing her fields converted into towns'. In 1929, Baldwin joined Ramsay MacDonald and David Lloyd George in endorsing CPRE's campaign for development to be 'directed with thoughtful and scrupulous attention to the charm of our land'. In conceding that with new methods of planning 'we ought to be able to effect necessary changes to avoid injuring a precious heritage,' England's political leaders were inviting an even more forceful CPRE push for legislation.

PLANNING TO PROTECT THE COUNTRYSIDE

One of MacDonald's first acts after winning the 1929 General Election was to offer CPRE his 'wholehearted support' in preserving the beauties which are 'threatened with desecration due to individual and commercial thoughtlessness'. Abercrombie and his colleagues soon began lobbying MacDonald on 'the wider use of town planning powers in rural areas', working closely with Edward Hilton-Young MP to draft the Rural Amenities Bill. The Bill was introduced to the Commons in February 1930 with the aim of giving County Councils control over development in rural areas for the first time, including through new powers to buy up threatened land. Hilton-Young praised CPRE's Westminster Hall exhibition for drawing attention to the 'deepest evil' of urbanisation: the 'lack of forethought in the development of the countryside . . . the higgledy-piggledy development in which building is foolishly scattered instead of being reasonably concentrated'.

Lady Cynthia Mosley, the MP for Stoke, hoped the Bill would address the sporadic development that could lead to 'the prodigal waste of a lovely tract of countryside that is utterly ruined by one glaring red roof', or 'some hideous bungalow put down at the wrong place. The time has come when we must definitely choose between the end of *laissez faire* or the end of rural England.' Clement Attlee added his support for the Bill, pointing out that 'there is a large part of our rural activities which ought to be removed from mere profiteering'. The novelist John Buchan MP observed that while the industrial magnates were quite prepared to industrialise the whole country, 'it is not the rich man who is going to suffer by the loss of rural amenities'.

LAISSEZ FAIRE'S LAST STAND

Despite its almost universally positive reception, the proposals in the Bill sent a shockwave through England's landowning elite. An angry correspondent to *The Times* was 'amazed to find members of the Conservative party giving their support to such Socialistic proposals . . . which cannot be justified on what are, after all, merely aesthetic grounds'. However, the mood of England was changing, with a Ministry of Health minute noting that 'systematic development has reached a stage beyond the control of private individuals'. *The Times* itself commented that the Bill was a valuable foundation for tackling the evils 'spreading with almost incredible rapidity', adding that 'others than Socialists are increasingly of opinion nowadays that the landowner is to some extent a trustee for the nation'.

An example of the kind of unplanned development that prompted CPRE's Westminster Hall exhibition of evils in 1930.

Bungaloid growth in its later stages. A twentieth century "Rural Slum" in Cheshire.
[Photo. C.P.R.E.]

The favourable reaction to the Bill meant that when the Labour Government fell, MacDonald's National Government carried over its provisions as the Town and Country Planning Bill. Just before the Town and Country Planning Act was passed in June 1932, Arthur Greenwood – recently succeeded by Hilton-Young as the responsible Minister – criticised a successful attempt to curtail the scope of the Act by 'a little knot of Diehard Tories'. When the 'diehards' declared they were actually Whigs, and asked Greenwood whom *he* represented, he replied: 'enlightened public opinion' of the kind inspired by groups 'like the Council for the Preservation of Rural England, which has done so much to promote a healthy interest in this subject'.

The rebels helped pass an amendment that meant the Act's protection would not apply to areas that were 'unlikely to be developed' or of little 'natural interest or beauty'. In the preceding debate, the Southwark MP, Sir Ian Horrobin, made a passionate plea for the protection of 'the ordinary normal quiet countryside', suggesting that picking out 'little pieces for protection' put the rest at even greater risk. Sir Stafford Cripps was sure that if the legislation did not cover all rural areas 'it is going to be absolutely fatal to country planning'.

Nevertheless, the amendment made it through to the Lords, to the dismay of the Earl of Listowel: 'Is not every inch of the English countryside untouched and unspoiled by the march of our industrial civilisation a thing of beauty and loveliness and a precious part of our

The Chilterns countryside showing the positive impact of the Rural Planning which Patrick Abercrombie hoped would balance 'existing features, natural and historic, and new growth' by respecting traditional patterns of rural land use and showing an 'appreciation of landscape beauty'.

national heritage?' Lord Mount Temple was confident the legislation would still ensure 'no more Peacehavens – monstrosities which are a disgrace to the twentieth century', urging that 'if we pass this Bill, the next generation will hold this Parliament in grateful remembrance'.

The journal of the Town Planning Institute felt that enacting such a radical piece of legislation during the Great Depression showed 'the recognition by the National Government that planning is essential to efficiency and economy'. However, the Act was far from an overnight success as the old habits of *laissez faire* died hard and industrialists and developers continued to wield considerable influence. CPRE soon published its 'To Plan or Not to Plan' pamphlet, urging local authorities to take up their new powers, and in 1934 Neville Chamberlain, ever the peacemaker, called for all sides to accept some impingement on 'personal interests and liberties'.

Despite the inevitable teething problems, the Ministry of Health was able to declare in 1936, that 'nobody who goes about the country today can fail to observe that the tide of sporadic, unregulated development is being stemmed, and that planning is beginning to leave a visible mark on the English countryside'. State loans for the preservation of public open spaces tripled to almost £2.5 million between 1932 and 1936, while the amount of England and Wales covered by planning schemes increased from 7 million to almost 20 million acres during the same period. Just ten years after he first made the case for it, Patrick Abercrombie's idea of Rural Planning was having a very real impact on the way we thought about the English countryside.

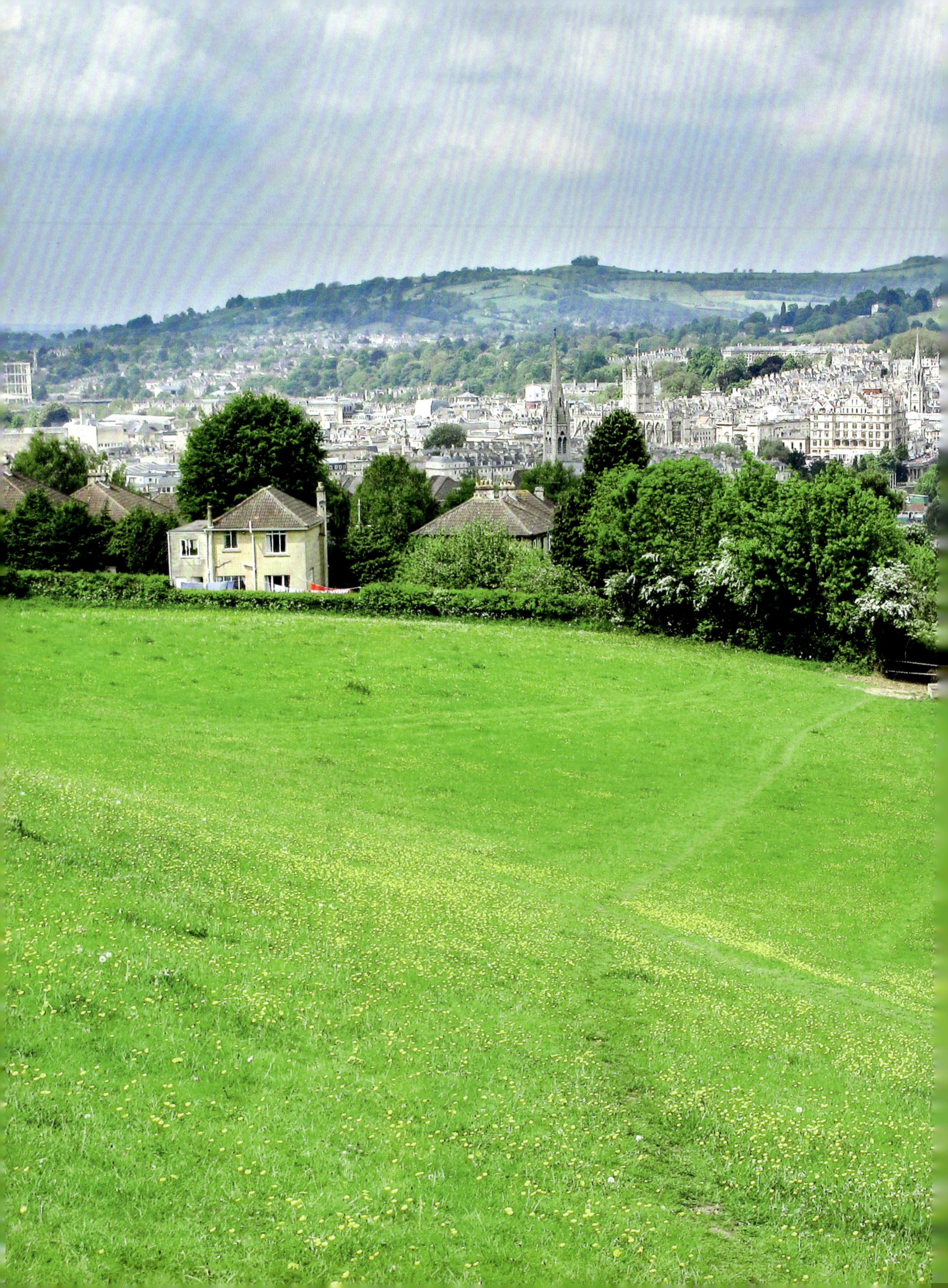

Idea 8

Democratic Planning

introduced by Max Hastings

There is an increasingly vocal lobby in Britain, demanding the partial or even total dismantling of the planning system, to promote a rapid increase in housebuilding, and also in the cause of economic growth. The Prime Minister and Chancellor of the Exchequer sometimes use rhetoric that suggests they share the view that planners inhibit national progress. Yet the history of the past century shows that planning has served us extraordinarily well in preserving the best of the countryside and controlling development. Look at the alternative, at the horrors that have unfolded in rural Ireland, for instance, where planning law is almost powerless to prevent all manner of reckless and tasteless building.

CPRE has played a critical role in shaping planning, and in fighting for the preservation of statutory restrictions on development. It is frightening to see how quickly great tracts of countryside can be lost, where the planners are overruled and the lords of steel and concrete gain sway. There will always be acute tensions between planners and builders, and of course the housing needs of future generations must be met. But we should recognise planners as heroes in the struggle to protect our landscape, as much today as in the twentieth century.

Although the 1932 Town and Country Planning Act marked the first sense of a collective national responsibility towards the countryside, the voluntary nature of the Act and the slow progress made by councils meant that planning was advancing across the countryside in a rather piecemeal manner. In 1936, Conservative MP Alfred Bossom argued that 'if there is to be real town and country planning we should not play with it; the whole subject must be considered on a nation wide plan'. Bossom seemed to have found an ally the following year when Harold Macmillan noted the 'farcical ineptitude' of the decision to build an aircraft factory on South East farmland when there was ready-made infrastructure and labour in the North.

If Britain's rush to re-arm was a case study in the need for national control over development, Macmillan argued in 1938's *The Middle Way* that such controls were essential for upholding democracy; efficient planning of the use of land and location of industry would mean the benefits of increased production could be 'directed towards raising the standard of comfort and security of all of the people'. Macmillan would no doubt have been encouraged by the previous year's comments by the Manchester industrialist Sir Ernest Simon, who had spoken of planning as part of 'a new spirit', which could create 'a reunified democracy determined to build a new civilisation'.

PLANNING FOR COMPLETE CONTROL?

In the run up to the Second World War, CPRE's Herbert Griffin lobbied civil servants for any national planning machinery to 'provide a proper balance between national, urban and rural points of view' knowing that, over the next few years, 'every inch of our beloved land shall be put to its most productive use'. Patrick Abercrombie, now CPRE's Chairman, looked even further ahead in 1940, making the case that a central planning ministry would be essential for post-war reconstruction. By 1942, Earl De La Warr went so far as to say that State control could provide a positive liberation from chaos through 'the planned development of a free people'; Lord Bledisloe cautioned that the State would need to be regulated by independent bodies like CPRE, which had already had to 'preserve the countryside against the somewhat impetuous wartime policies of different Government Departments'.

When Winston Churchill's minister without portfolio Sir William Jowitt introduced a Bill to create a Ministry of Town and Country Planning in January 1943, Bledisloe must have been worried that the preservation of countryside did not appear in the list of what 'Planning with a big P' was expected to encompass: 'the location of industry, the prevention of cyclical depressions, education, public health, social services, agricultural policy, the development of roads, harbours and ports, and last, but by no means least, financial policy'. Of course, planning still remained the obvious way to deal with the threats to England's countryside, and the war only served to heighten the sense of threat.

Winston Churchill played a huge role in ensuring there was still an England left to be planned, but did not share the belief in the need to plan

PAGES 74–5
Famed for its architecture and quality of life – and surrounded by protected and accessible landscapes – Bath is a classic example of the benefits of Democratic Planning.

IDEA 8 • DEMOCRATIC PLANNING

Patrick Abercrombie's 1945 plan for Ongar New Town drawn by his godson Peter Shepheard (a future President of the Landscape Institute); featuring the 'broad vistas' feared by Churchill, the plan also includes 'green corridors' to bring the surrounding countryside into the town.

its reconstruction. For Churchill, planning was not a redeeming feature of dictatorships; it was the cause of them. Towards the end of the war he even mocked his cabinet's enthusiasm for planning 'broad vistas', saying 'give me the eighteenth-century alley where the harlot plies her trade, and none of this new fangled planning doctrine'. A strange point of view perhaps, in light of his comments to his wife in the dark days of May 1940, soon after becoming Prime Minister; as they stood on the terrace at Chartwell looking out over the Weald of Kent, he told her: 'That is what we are fighting for, Clemmie.'

This failure to understand how planning could help save the countryside partly explains why Churchill was not returned to Downing Street, despite the eternal gratitude of the nation. He misjudged the mood of the public, siding with certain 'Colonel Blimp' types who argued that the very essence of 'Englishness' was the result of the traditionally unplanned nature of development in England. Major Maurice Petherick MP led the vocal minority in warning that planning controls represented 'as much interference with the liberty of the subject as the unfortunate subject can swallow without revolution'. But this vision of Englishness was one that respected the freedom of the individual to make a mess more that the rights of the communities who had to live with the results.

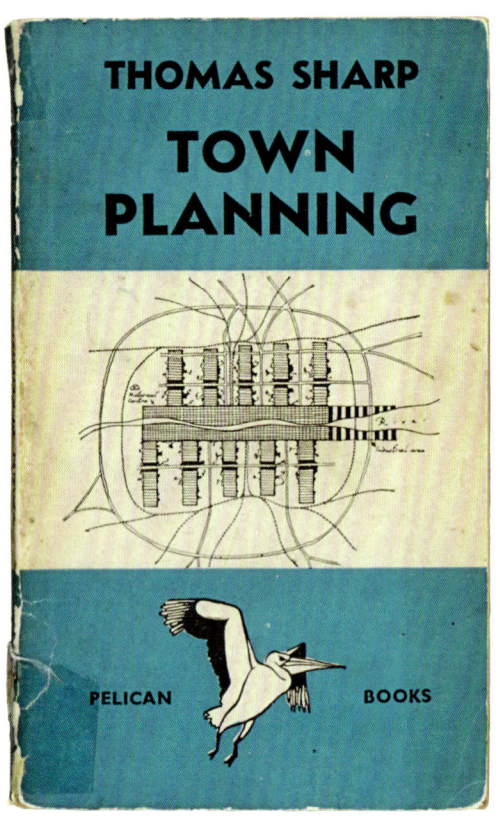

Thomas Sharp's 1940 Pelican on *Town Planning* had been devoured by 250,000 readers enthused by the idea that planning would not only preserve 'our physical environment', it would also 'save and fulfil democracy itself'. So while Churchill's old guard felt planning might be the first dangerous step towards totalitarianism – citing the grand planning schemes of Nazi Germany, Fascist Italy and Communist Russia – the planning system envisaged by Sharp and Abercrombie was always intended to be fully democratic: local authorities having the freedom to plan what happened in their areas, while being accountable to local people. Sharp felt that Holland's reclamation of the Zuiderzee and the great public works of the United States proved that democracies could also 'plan ahead and undertake heroic large-visioned works', concluding his book with these lines:

> The political ingenuity of a nation which for long centuries led the world along the path of just and progressive democratic government is surely capable of evolving some form of organisation that will lead to the satisfaction of that most necessary of democratic ideals, the planning of the environment of human living to achieve happiness, efficiency and beauty.

Thomas Sharp's 1940 Pelican summed up the growing feeling that, while a 'democratic muddle' was preferable to a dictatorship, 'the English muddle is nevertheless a matter for shame... haphazard, short-term, and small-visioned'.

POST-WAR OPTIMISM FOR PLANNING

A draft Ministry of Town and Country Planning circular of 1945 gives a sense of the positive post-war ambitions for planning in its closing words: 'It is for the planning authorities, as the nation's prime and statutory trustees, to safeguard the beauty of our country for the physical and spiritual refreshment of present and future generations.' Yet the 1932 Act meant that the same authorities could only safeguard this beauty if they could afford to compensate landowners for refusing a development application. A new Town and Country Planning Bill proposed that landowners should not gain from the appreciation in land value which planning permission brought with it; nor should local authorities be compromised by the need to pay compensation which could be paid from State coffers. Planning Minister Lewis Silkin hoped these changes would usher in 'an era in which human happiness, beauty, and culture will play a greater part in its social and economic life than they have ever done before'.

The eventual Town and Country Planning Act was passed in 1947, 'effectively nationalising development rights in land' as former CPRE Planning Officer Marion Shoard put it forty years later. Post-1947, landowners had to apply for the right to build from a democratically elected local planning authority. CPRE even convinced the Government to allow the public examination and reporting of cases where developers appealed against rejected applications. When the Act came into force in July 1948, they rejoiced that after the free-for-all of *laissez faire*, now 'the ownership of land carries with it little more than the bare right to go on using it for its existing purpose'. The right of the individual to do as they liked could no longer outweigh the rights of local communities to have a say in decisions that could destroy their green spaces.

The loose 'planning schemes' created by the 1932 Act were swept aside by comprehensive Development Plans, which would be created with the involvement of local people. The plans would not just give great detail in how land would be used for industry, commerce, housing, and infrastructure, but, most importantly for CPRE, would allocate land to be reserved for agriculture and open spaces. CPRE believed the Act would help 'produce something new and beautiful' while maintaining the 'integrity of the English countryside'. Not everyone was so enthusiastic about the Act; when Churchill returned to 10 Downing Street in 1951, one of his first acts was to erase his hated word 'planning', renaming the Ministry of Planning as the Ministry of Housing and Local Government. Churchill insisted that the change was made 'in order to emphasise the importance of housing', and that the word 'planning' was omitted 'for reasons of brevity and not of policy'.

However, Churchill's Housing Minister was that old ally of planning, Harold Macmillan. In December 1952, Macmillan said: 'Planning in its broad sense has come to stay – to preserve good agricultural land; to encourage the development we want in the proper places; to restrain the interwar sprawl of the growing cities and to preserve the countryside.' Despite these fine words, the Government hinted that the Treasury would not be prepared to foot the bill for compensation where councils refused development. CPRE suspected a lack of support for planning from the Prime Minister would mean that 'at times of financial stress, the interests of amenity and good planning may be overridden by over-rigorous Treasury control'. Tellingly, Churchill wanted to reduce the cost of planning compensation from £300 million (to cover the first five years of claims) to £100 million 'a generation'.

In October 1953, many of CPRE's suspicions were confirmed when the Ministry released a circular aimed at reducing the number of planning appeals by giving applicants 'the benefit of the doubt'. They were only partially reassured by Macmillan's speech to the CPRE AGM in 1954, in which he warned that 'if, in a period of national stringency good planning is sacrificed merely on financial grounds, it will probably never be regained'. Macmillan finished by acknowledging 'the main battle of CPRE has been won after a very long fight. There is a general acceptance that in so small an island one can not allow the complete individual freedom which might have been possible in more primitive days.'

'Town and country planning is now the single most important factor affecting the look of Britain. And we meddle with it at our peril!'

A LEGACY OF DEMOCRACY

Tristram Hunt has said that the planning system created in 1947 was a 'legislative monument to the work of CPRE' in defeating *laissez faire* sprawl'. And, by and large, it has continued to be an effective defender of the countryside despite the constant pressure to chip away at its protection, most notably during the return to *laissez faire* values in planning in the 1980s. In 2012, the Coalition Government's 'National Planning Policy Framework' attempted to boost the flagging economy by replacing the Act's presumption against building on the countryside with what campaigners called a 'developers' charter' designed – despite the denials of ministers – to encourage profitable development on England's green fields.

Despite these attacks, Democratic Planning has done a difficult job pretty well. It has historically delivered the housing, industry and infrastructure we need while preserving an incredible amount of our countryside. The archaeologist Francis Pryor says the fact that Britain has somehow managed to 'retain its uncluttered rural areas is almost entirely down to planning. Town and country planning is now the single most important factor affecting the look of Britain. And we meddle with it at our peril!' The former Chairman of the National Trust, Simon Jenkins, has argued that the measures of the 1947 Act gave 'half a century of primacy' to 'distinguishing woods, commons and farmland from built-up areas'. This protection, he argues, has left a landscape that belies England's status as the sixth most crowded country in the world: 'firm planning control has mostly prevented the sprawling municipalities familiar elsewhere in the world. Planning is an English success.'

Britain's leading post-war planning academic, Sir Peter Hall, ranked the 1947 Act as 'one of the great milestones of the Attlee Government, together with the independence of India and the NHS'. For Professor Hall, the most important feature of the Act was that it 'allowed local authorities to say no, an absolute no, to proposals for development, with no fear of paying astronomic sums in compensation for lost development rights'. Marion Shoard has said that 'without this system of development control, the face of Britain would have been ravaged by building'. We do not always appreciate its more impenetrable technicalities, but the planning system established by the 1947 Act means we all have a democratic right to know what will be built near us, and a voice to do something about it.

OPPOSITE
Towns like Totnes in Devon show how Democratic Planning has helped protect our heritage while maintaining the distinction between farmland and built-up areas.

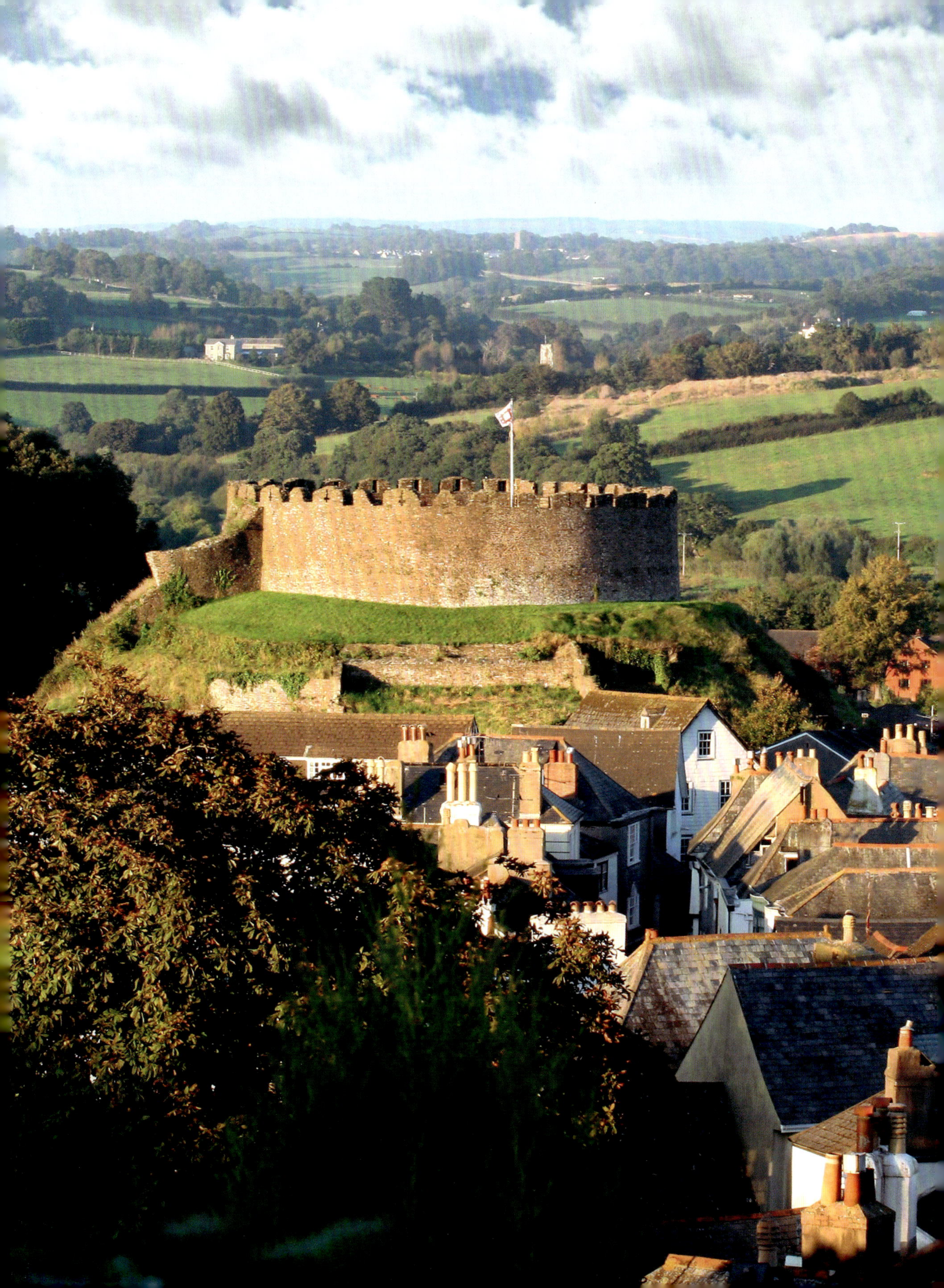

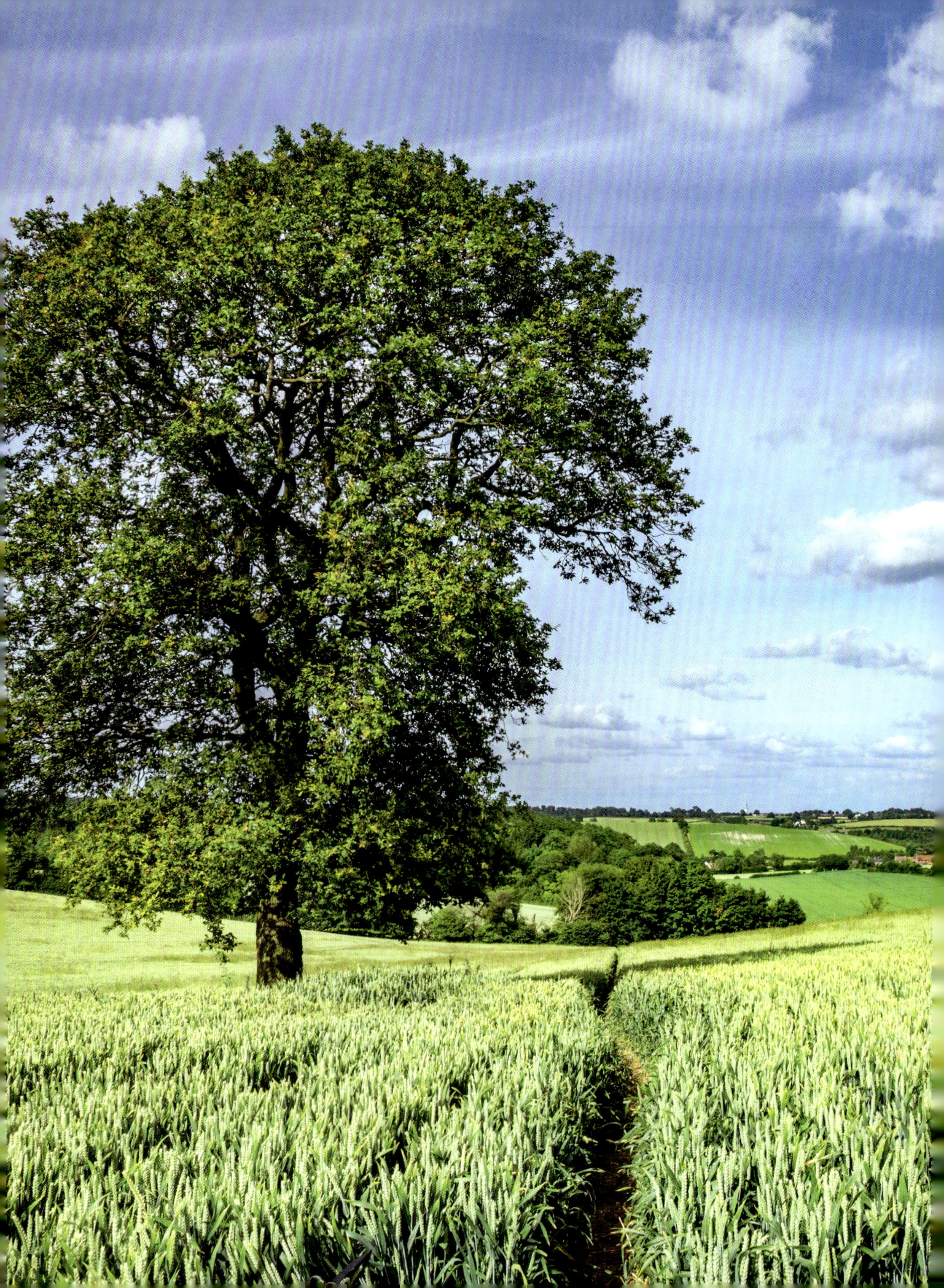

Idea 9

Green Belts

introduced by Ray Mears

The Green Belt, an idea which is still less than 100 years old, was conceived of to prevent urban sprawl, and it's proved to be one of the most successful acts in the history of conservation. These buffers are maturing into internationally important habitats, often richly diverse in species. They also create healthier air, and make our towns and cities happier, more relaxing places to live. It's really important that we preserve the Green Belt. The moment you start to dig at it, you put a hole in the dyke, and you can't put that back. I think it was a really bright idea when it was put forward – one of those great acts of politics. If anything, we should be trying to extend the Green Belt.

Green Belts are an icon of England, envied and emulated around the world. Fourteen of our cities and major conurbations are encircled by a protected belt of green open land containing farms, allotments, Country Parks, Community Forests and countless other precious landscapes designated for their outstanding beauty or scientific interest. Green Belts are so popular, and so staunchly defended from development, because they offer recreation and escape for the 60 per cent of the population who live within them; they truly are the nation's 'countryside next door'. Most of us who live in towns refer to the countryside outside them as Green Belt, regardless of whether it is officially designated.

When Green Belts became official Government policy in 1955, they were not designed to protect land for its beauty or recreational value, or even its value as farmland. They were solely designed to stop urban sprawl by checking further growth of urban areas, preventing them from merging together, and preserving their landscape setting. In his 2008 speech to their AGM, the then CPRE President Bill Bryson, noted that Green Belts 'remain the crucial instrument for containing urban sprawl and driving regeneration in our towns and cities'. This functional concept has captured the English imagination so firmly because its main side effect has been to protect the countryside closest to our urban areas.

ANCIENT ORIGINS

The concept of Green Belts gained real influence in England only in the early twentieth century, but the idea that towns should benefit from their surrounding countryside has its origins in the Old Testament. The Book of Numbers was compiled in the fifth century BC and states: 'Command the Israelites to give the Levites towns to live in from the inheritance the Israelites will possess. And give them pasturelands around the towns.' But, of course, the size of cities was governed by even more pressing concerns than the desire for open space. From the very earliest cities of ancient Mesopotamia, up to the walled cities of Roman England, settlements traditionally had clearly defined boundaries borne out of the need for fortifications to keep out invaders.

It was the ancient Greeks – who else? – who added a social component. In the middle of the fourth century BC, Aristotle's *Politics* argued that it was vital that cities remained small enough for 'the citizens to know one another'. But conviviality was not the only aim. Aristotle believed that limiting a city to a population of 30,000 would mean its citizens could be fed from the surrounding agricultural land, reducing the need for the food miles, which, even then, compromised the cost, freshness and reliability of imported produce.

Here in England, Elizabeth I attempted to halt the practice of building outside the city walls with a proclamation of 1580, designed to stop the spread of disease. This was followed by a remarkably concise Act of 1593, which seemed to be the first conscious attempt to limit the spread of London, specifying that 'commons lying within three miles of London shall not be

PAGES 82–3
A footpath through farmland in the Metropolitan Green Belt in Hertfordshire, proving that Green Belts do far more than simply stop urban sprawl.

The importance of maintaining a balance between town and country is made clear in Ambrogio Lorenzetti's 1339 frescoes on the Effects of Good and Bad Government *in Siena's town hall; the 'good' is represented by people moving, effortlessly, between a vibrant city and the beautiful and bountiful countryside surrounding it.*

inclosed'. Elizabeth's attempts ultimately failed because, as *The Times* explained in 1936, 'for those who wanted to build in the prescribed zone and could afford to pay for them, "permits" were available'. From just 250,000 in 1601, London had reached a million inhabitants by 1801, at which point it began adding the population of late-Elizabethan London every decade.

CREATING A BETTER SOCIETY

The campaign to preserve London's commons in the mid-nineteenth century had been fought following the original biblical ideal of providing cities with access to the vital amenity of open green space. The passion of the commons campaigners was complemented by the practical skills of a new generation of urban thinkers. In 1835, Colonel William Light's plan for Adelaide incorporated a ring of 1,700 acres of parkland between the city and its suburbs. Australasia had become a hotbed of urban experimentation, as England's finest surveyors were given free rein to implement their visions of the ideal city from a completely blank canvas.

Colonel Light had been inspired by the pioneering colonial administrator Edward Gibbon Wakefield; a King's Messenger during the Battle of Waterloo, Wakefield had pondered the causes of urban overcrowding in England during his incarceration in Newgate Gaol for abducting a fifteen-year-old heiress. He resolved to build new cities in the colonies to relieve population pressure at home and to create exemplars of urban design, founding the National Colonisation Society upon his release. The economist John Stuart Mill was a member of the Society, and his *Principles of Political Economy* of 1848 followed Aristotle's logic, arguing that a town should not expand beyond the capacity of the surrounding farms to feed it. The following year, the former Sheffield MP James Silk Buckingham planned 'Victoria', a model town for 10,000 people within a square mile surrounded by a belt of 'farmland 10,000 acres broad'.

A long-time social-reformer, Buckingham's plan placed the 'residences of the working classes nearest the green fields on the immediate edge of town', where they would be 'favourable to their health'. Inspired by Sir Christopher Wren's plan for the rebuilding of post-Great Fire London, Buckingham recommended that streets should 'open directly on the surrounding country' so that 'the pure atmosphere of the fields will rush directly through the town, attenuating the noxious gases and revivifying the used air'.

The city of Birmingham is barely visible in the distance from the Clent Hills Country Park, a popular leisure destination in the West Midlands Green Belt.

UTOPIAN VISIONS VERSUS LOATHSOME REALITY

Charles Dickens' 'Coketown' – the fictional setting of *Hard Times* (published in 1854, five years after the plan for 'Victoria') – helped popularise the idea that something on the lines of Buckingham's proposal was needed to 'revivify' the sprawling towns, 'out of which interminable serpents of smoke trailed themselves for ever and ever, and never got uncoiled'. In 1868, Ruskin described how the pursuit of beauty could create the perfect city in *The Mystery of Life and its Arts*:

> . . . the building of more [houses], beautifully, and in groups of limited extent, [should be] kept in proportion to their streams, and walled round, so that there may be no festering and wretched suburb, but the open country, with a belt of beautiful garden and orchard round the walls, so that from any part of the city perfectly fresh air and grass, and sight of far horizon, might be reachable in a few minutes' walk.

William Morris echoed his mentor's concerns – and anticipated his own *News From Nowhere* – two years later, in a letter to a friend from 'loathsome London'; Morris dreamt that people could live in 'communities among gardens and green fields, so that you could be in the countryside in five minutes' walk'. In the same year, Ruskin's Oxford 'Lectures on Art' recounted a clergyman's confession that he could only enter London with closed eyes, 'less the sight of the blocks of houses which the railroad intersected in the suburbs should unfit him, by the horror of it, for the day's work'. Ruskin went on to claim that 'it is not possible to have any right morality, happiness or art, in any country where cities are thus built, or rather, clotted and coagulated; spots of a dreadful mildew, spreading by patches and blotches over the country they consume'. Though his vision for a compact, walkable town 'girded with garlands of gardens' was utterly at odds with the prevailing forces of the Industrial Revolution, Ruskin resolved to 'have nothing to do with its possibility, but only with its indispensability'.

BELTS AND GIRDLES: NEW IDEAS FOR LONDON

In 1875, Octavia Hill seized on Ruskin's ideas in calling for Swiss Cottage Fields to be preserved as part of a continuous 'green belt' of parkland around London, a theme revisited in her *More Air for London* of 1888, which may have been prompted by Ruskin's letter to her warning that: 'London is as utterly doomed as Gomorrah . . . I have to labour wholly to fence round fresh fields beyond the smoke of the torment.' Remarkably, Hill's 'green belt' was not heard of again for nearly fifty years, being swiftly superseded by 'green girdle', a phrase that seemed to suggest a more restrictive 'fastening in' of the city. The 'girdle' concept was made fashionable by Cologne's *Grüngürtel* reservation of open green space around the city in the late-nineteenth century. A similar scheme in Vienna became formally protected by a 1905 law to preserve the 'green girdle' for public recreation.

In England, the phrase was popularised by Reginald Barabazon, the Earl of Meath, who recommended that the London County Council link existing parks in a continuous 'green girdle'. Meath had been inspired by the Louisville 'parkway' of tree-lined avenues linking wedges of countryside, designed to bring the landscape into the city and provide a 'poetic and tranquilising influence' on residents. The parkway's creator Frederick Olmsted was working on a similar project for Boston named the 'Emerald Necklace'; it is possible that only Meath's preference for the tougher-sounding 'girdle' means that we do not now talk about protecting England's emerald necklaces!

While Meath continued to argue that London's suburban parks could be linked 'by broad sylvan avenues and approaches', Hammersmith MP William Bull presented a more radical plan in *The Sphere* illustrated newspaper in 1901. Bull proposed 'a circle of green sward or trees which would remain permanently inviolate' and provide 'the best memorial to Queen Victoria'.

Ultimately, neither Bull nor Meath gained widespread support for their plans, and it was another reformer who would become more influential in the early years of the twentieth century. Bull and Meath had failed to address the inevitable fact that if towns were to stop sprawling, the growing population would have to be housed somewhere, but in 1898, Ebenezer Howard made the case for new 'garden cities' to help disperse the urban population. Designed to house 32,000 people on 1,000 acres of land, each of Howard's garden cities would be surrounded by 5,000 acres of 'permanent agricultural reservation'.

'The crying need of today is that we should step in and save what is still rural from suburbanisation.'

The idea of garden cities became reality in 1903 with the development of Letchworth, 34 miles outside London. The problem of keeping house prices low enough meant that it was not seen as a lasting solution to London's sprawl, becoming more of a model for interwar suburbia than a shining example of how to build new communities. In 1901, H.G. Wells had taken the garden city ideal of low-density homes with gardens to its logical conclusion; his *Anticipations* of the year 2000 foresaw the whole country transformed into one big leafy suburb, defined as 'Town-Country' and retaining the best attributes of both. The incredulous planner Thomas Sharp later re-christened Wells' hybrid as 'Neither-Town-Nor-Country'.

THE MOVEMENT GAINS MOMENTUM

As the sprawl of the metropolis continued unabated, a succession of prominent men recommended variants of the 'green girdle' for the fastest growing city of all: the architect Paul Waterhouse showed a ring of 'park, woodland and playing fields' in his *Imaginary Plan for London of 1907*; in 1909 Raymond Unwin, the town planner and former secretary of William Morris' branch of the Socialist League, suggested 'some intervening belt of park or agricultural land' to prevent neighbouring towns joining up; the following year another respected town planner, George Pepler, proposed a 'parkway' road within a green reservation circling a 10-mile radius from Charing Cross; and, in 1913, the London Society prepared a 'Development Plan of Greater London' which suggested preserving and connecting 'strips of land bordering the numerous streams in the London area'.

So many ideas – and yet still no action. The notion that town and country should remain distinct sounded old-fashioned when suburbia promised the best of both worlds and, rebranded as *Metro-land*, had some of the most powerful marketing ever seen in England from 1915. Furthermore, it was difficult for campaigners and planners to compete with the forces of commerce without financial backing. But thanks to the generosity of landowners and the foresight of the Corporation of London, many treasured green spaces on the edge of London had already been saved in the belief they might one day form part of a greater whole. As early as 1869, opposition to the encroachment of Coulsdon Common eventually led to the donation to the Corporation of over 400 acres of land. Nearby Banstead Downs was protected in the same way soon after, together with Epping Forest and Burnham Beeches.

IDEA 9 • GREEN BELTS

Clive Gardiner's 1934 poster showed Epping Forest 'at London's service', thanks to the foresight of the City in buying it up almost fifty years earlier. The forest would become part of the Metropolitan Green Belt in 1938.

When, in 1924, the London County Council resolved to create a 'green belt' as 'an inviolable rural zone around London', it was the first use of the phrase 'green belt' since Octavia Hill forty-nine years earlier. Fittingly, the permanence suggested by 'inviolable' evoked the 'inalienable' ownership of land by Hill's National Trust. In 1927, the Minister of Health Neville Chamberlain asked the Greater London Regional Planning Committee to consider providing London with 'an agricultural belt'. The following year, CPRE proposed the introduction of 'zoning restrictions' to allow local authorities to 'reserve areas permanently as agricultural land or open spaces'. Though a 1929 Parliamentary debate saw the Government pressed to help local authorities buy up land for a London green belt, the impact of the Wall Street Crash soon made this financially impractical.

RESERVING LAND FOR GREEN BELTS

Progress was finally made in 1932, but not in London; CPRE's Sheffield and Peak District branch helped negotiate the purchase of 448 acres on the outskirts of Sheffield, and over the next decade worked with philanthropist Alderman Graves to acquire over a thousand more to prevent the spread of the city. Then, in June 1933, *The Times* reported on the Oxford Preservation Trust's acquisition of parkland outside the city: 'Except at one or two points Oxford is still surrounded by a girdle of green, and South Park, Headington was now a central part of the green belt of Oxford.' Writer and MP John Buchan articulated some of the universal benefits of Green Belts by pointing out that 'Oxford depended for its charm very largely on its environment', adding that 'if they spoilt the setting they would take away much of the charm of the jewel'. Birmingham soon joined in as the Cadbury family secured land for a Green Belt in co-operation with the City Council and the National Trust.

The London County Council responded in January 1935, with its leader Herbert Morrison announcing the creation of a £2 million fund to help counties bordering London acquire and reserve land as part of a Green Belt. *The Times* commissioned special photographic articles documenting the progress of the Green Belt, but warning that 'something else has been progressing much faster: and that is what is now called development – in plain English, building over land not built on before'. CPRE argued for legislation to encourage Green Belts around towns without Government funds or private reservations: 'the crying need of today is that we should step in and save what is still rural from suburbanisation.'

CPRE's National Conference of October 1937 became a conference on the future of Green Belts, with Abercrombie in the Chair asserting

89

that even 'small towns should think seriously about . . . agricultural reservations' on their perimeter, which should cover a wide area 'to be really effective; that is the difference between a Green Belt and a green ribbon'. Conference delegates were given a tour of the Birmingham Green Belt by its founder George Cadbury: 'People are getting educated to know what a Green Belt is and what it looks like,' Cadbury explained. 'The idea is taking root.' In June 1938, the Green Belt (London and Home Counties) Act was finally passed, meaning that land already reserved by county councils would be entered into covenants, preventing its future sale and any new building. The Metropolitan Green Belt was on the map, and although CPRE's President, Lord Crawford, told the London Society it was often 'greener on the map than in the field', he was adamant that 'however unattractive, it is the potential bulwark against London's continued expansion'.

CELEBRATING A REMARKABLE ACHIEVEMENT
The Second World War may have interrupted further progress on Green Belts, but not before the Metropolitan Green Belt Act could be assessed in its historic context. For someone intimately associated with the National Trust, the historian G.M. Trevelyan could pay no greater tribute than remarking that while the Trust had saved 70,000 acres in forty years, the Green Belt scheme had done the same in less than four. He praised the efforts of local authorities as a future model for conservation:

> Really something big has at last been done on the part of public bodies out of public funds. The National Trust, even in collaboration with CPRE is not going to save the beauty of England. You have got to get the corporations going and the nation going.

Lord Crawford called the Green Belt 'a great ideal . . . even if we had been rather late in the day to realise it', while Herbert Morrison hailed 'one of the most rapid achievements in local Government'. In April 1940, Morrison even claimed kinship with Elizabeth I as a fellow crusader against sprawl:

> The trouble about Queen Elizabeth's effort was that while she got legislation passed, the Conservative anarchists of that time defied authority, and it never came off. It is a great pity that I did not live in the days of Queen Elizabeth, because she and I together would have done something about it.

At the same time, Thomas Sharp was popularising the idea that the 'emphatic physical distinction' between town and country offered by Green Belts would preserve the thrill of passing from one to the other: 'one of

Julius Kupfer-Sachs' 1939 illustration of the runaway sprawl of London after 1889. The map shows why there was such powerful motivation for the creation of the Metropolitan Green Belt in 1938, and how effective it has been in protecting the surrounding countryside ever since.

the oldest and most fundamental excitements that civilised man has known'. Commissioned to shape the post-war reconstruction, Abercrombie's *Greater London Plan* of 1944 reinforced the Green Belt as 'of paramount importance to London'. His zonal plan incorporated a 'Green Belt ring', 'to provide recreation and fresh food for the Londoner, and to prevent further continuous suburban outward growth'. Abercrombie's plan earned him a knighthood and was conceived in the most remarkable circumstances; after an air raid had seriously damaged his offices, Abercrombie defied wardens to retrieve a case of vintage claret from the cellar, which he promptly shared with his staff, boosting their morale before they started sorting through the debris.

The landmark Town and Country Planning Act of 1947 allowed local authorities to safeguard land for Green Belt proposals in the newly required Development Plans without having to buy it themselves. However, by 1952, Abercrombie bemoaned the lack of progress on provincial Green Belts; three more years of CPRE pressure finally forced Housing Minister Duncan Sandys to issue a momentous Circular in August 1955, which urged all urban councils to engage in the 'formal designation of clearly defined green belts'. Marking the first Government acknowledgement of the national importance of Green Belts, Sandys stressed that 'for the well-being of our people and for the preservation of the countryside, we have a clear duty to do all we can to prevent the further unrestricted sprawl of the great cities'.

FROM NATIONAL POLICY TO NATIONAL ICON

After a campaign spanning generations, Green Belts were finally given State support, and have rarely looked back since. Immediately following the Circular, CPRE branches across the country began devising Green Belt plans for their towns and cities, while over the decades, the total area of England's Green Belts has continually increased as more and more of our countryside became designated. In fact, the amount of England covered by Green Belt protection has doubled since 1978.

And yet, with the economic downturn increasing the temptation to see the expansion of our towns and cities as a cure-all, the pressure on Green Belts in 2016 is as intense as it has been since the recession of the early 1980s. Then, CPRE fought off a test-case proposal for a new settlement at Tillingham Hall in the Essex Green Belt and defeated two Government Circulars which would have completely undermined Sandys' of 1955; more recently, they have forced a Government retreat from talk of loosening the Green Belt to David Cameron's 2015 statement that 'the line remains scored in the sand – that land is precious. Protecting the Green Belt is paramount.' But with the Government's own 2015 statistics showing we are losing Green Belt land at the fastest rate since records began in 1997, campaigners are calling for the Government to turn its rhetoric into action.

There is certainly no shortage of popular support for the Green Belt: a 2014 *Guardian* survey showed that 75 per cent of the public are opposed to building on Green Belt as a solution to the housing crisis, while CPRE's major 2010 survey with Natural England on the state of the Green Belt showed what a crucial asset it is to the nation. England's fourteen Green Belts cover over 13 per cent of the country, providing open countryside for the 30 million people who live in the urban areas within them. Green Belts were never designated for their landscape quality, and are constantly attacked by opponents for their supposed 'scruffiness', and yet 95 per cent of people value their beauty. Moreover, less than 0.2 per cent of Green Belt land is vacant or derelict, and 9 per cent of Green Belt land is designated in Areas of Outstanding Natural Beauty.

The 6,330 square miles of England's Green Belts continue to provide rich and varied opportunities for recreation, containing, as they do, 44 per cent of our Country Parks, 33 per cent of our Local Nature Reserves, 17 per cent of our Public Rights of Way and 27 per cent of our National Cycle Routes. The natural habitats provided by Green Belt land include 41 per cent of our Community Forests, 19 per cent of our ancient woodland, and 221,000 acres of Sites of Special Scientific Interest. Of course, England's Green Belts still provide the food security that has been an issue for virtually every one of their supporters from Aristotle to Abercrombie; 66 per cent of the Green Belt is farmed, and it contains the same proportion of Grade 1 and 2 agricultural land as the rest of England. And yet, there is so much potential to 'green the Green Belt' and make it an even more valuable community asset, with the creation of new forests, footpaths and facilities to help local people learn about nature and grow their own food.

The magnificent natural resource of our Green Belts takes on even greater significance in the fight against climate change. They provide us with the trees, hedgerows and fields, which store carbon, absorb floodwater and reduce the 'heat island' effect of big cities. Of course, the by-product of preventing sprawl is that we have saved priceless areas of green space on urban doorsteps. Darwin's 'landscape laboratory' – the Metropolitan Green Belt around Downe, near Orpington – has been considered for World Heritage status in recognition of the importance of its countryside. Londoners can enjoy the unchanged Green Belt landscapes studied by Darwin 150 years ago, just minutes from the southern edge of the city. Elizabeth I and her band of remarkable followers would be proud indeed.

OPPOSITE

Green Belts are constantly attacked because of the supposed scruffiness of the land on the 'urban fringe'. This photo of the Metropolitan Green Belt in Hertfordshire is a more accurate representation, showing why around 95 per cent of people value their beauty.

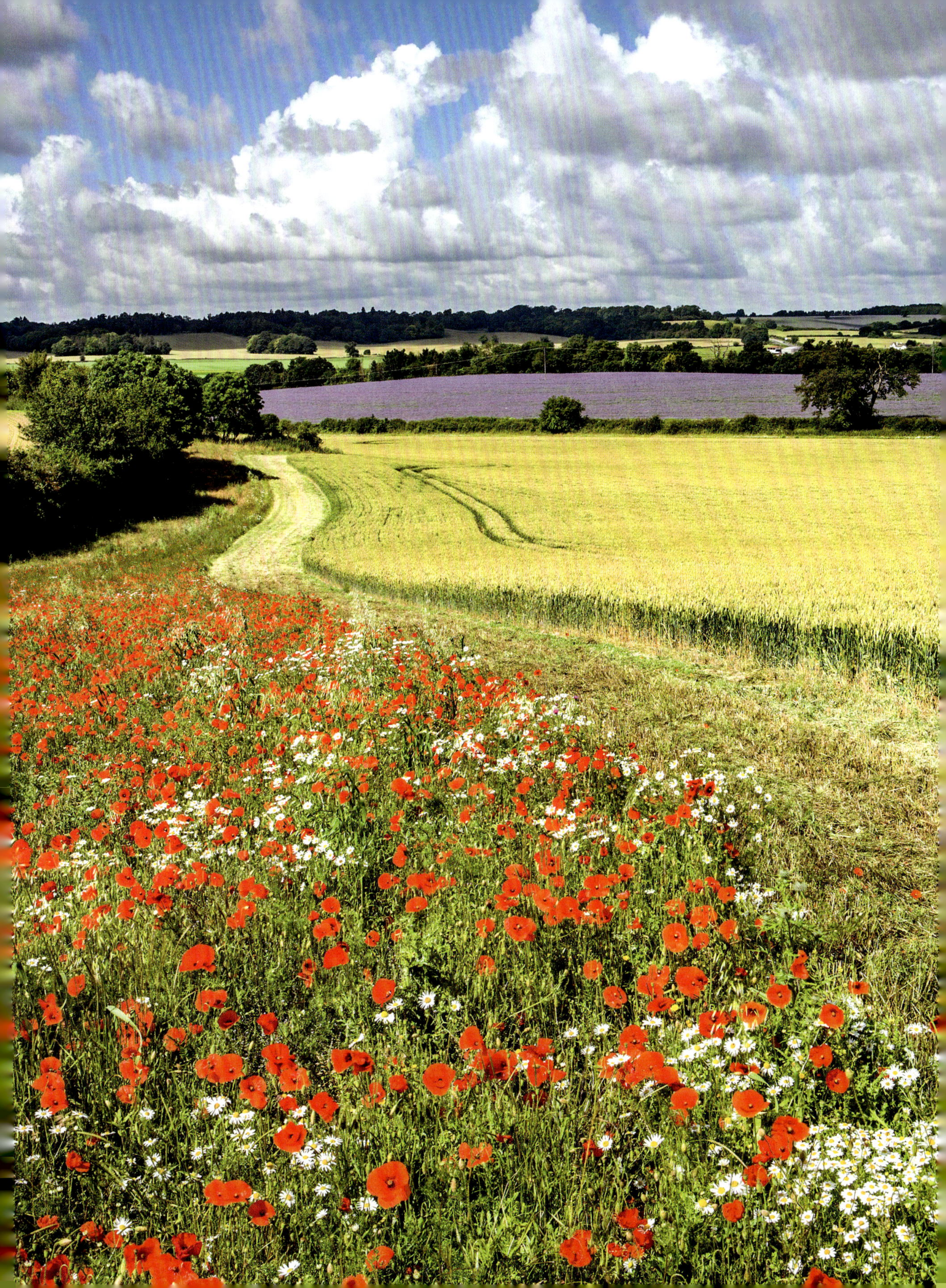

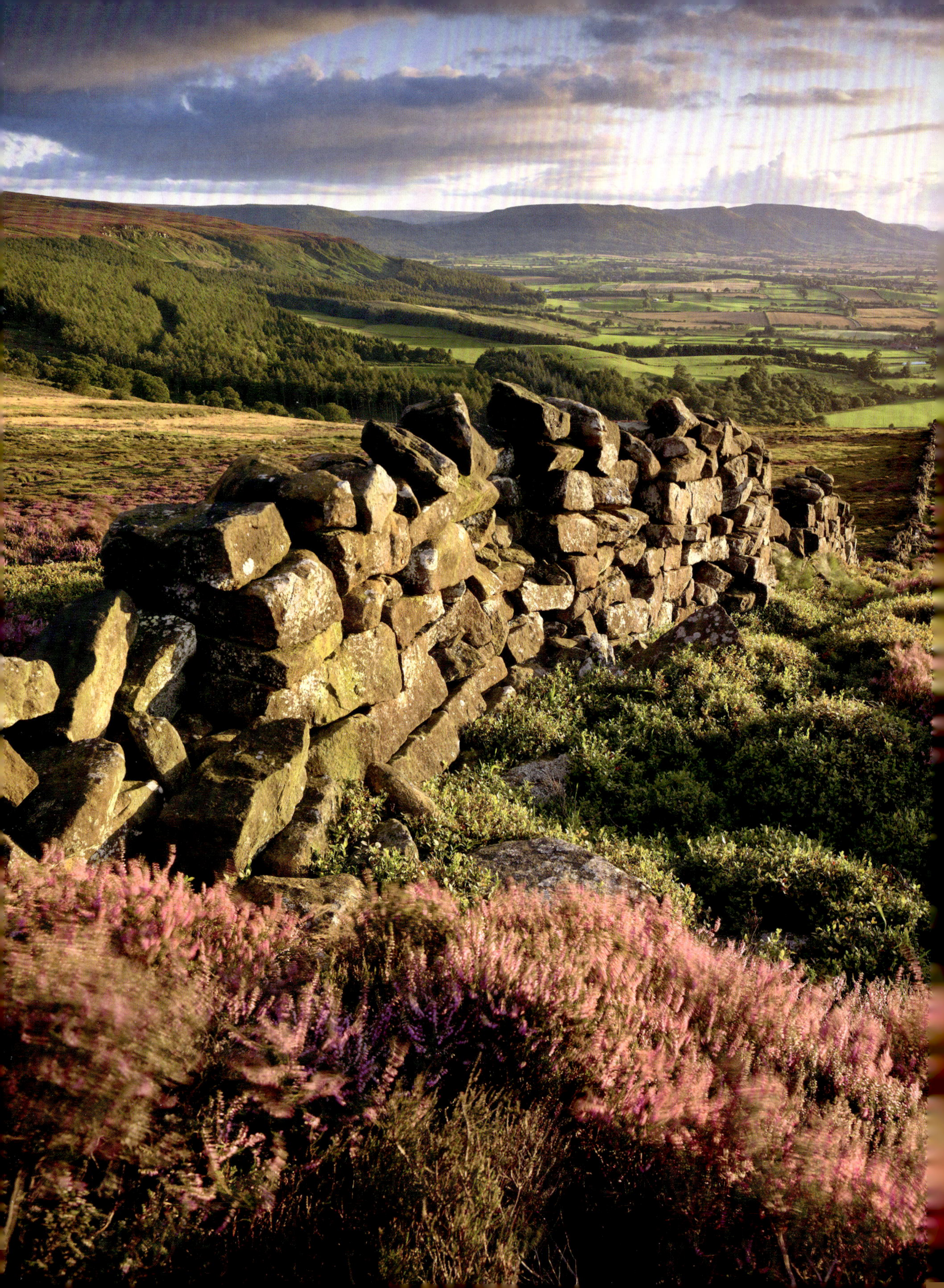

Idea 10

National Parks

introduced by Satish Kumar

The National Parks have extrinsic as well as intrinsic value. The wild, the mysterious, the magical and the inspiring landscapes, biodiversity and beauty of these natural treasures maintain the health of people and other forms of life. In their profound wisdom and foresight, our ancestors created these wonderful parks so that people could celebrate life and appreciate the utter interdependence of humanity and the natural world. National Parks – with their gracious, rugged and romantic quality – provide places where they can be totally united. CPRE, together with many other organisations and individuals upholding the value of untamed natural habitat, are to be congratulated for their commitment to protect and preserve these National Parks and other landscapes of natural integrity.

It is impossible to look beyond Wordsworth's *Guide to the Lakes* of 1835 as the origin of National Parks in England. The great poet famously described the Lakes as 'a sort of national property in which every man has a right and interest who has an eye to perceive and a heart to enjoy'. It is hard to imagine now what an extraordinary statement that would have been at the time. The idea that parts of the countryside could be 'owned' by the nation predates even the National Trust by seventy-five years.

Over in the New World, the vast size of the country and the inhospitable terrain of the wildest parts gave America's nature philosophers the freedom to be even more daring. Ralph Waldo Emerson, in his *Nature* of 1836, opined that 'the landscape belongs to the person who looks at it'. In 1841, the renowned painter of native American Indians George Catlin – fearing the loss of wild plains which were the habitats of both buffalo and Indian – envisaged 'a Nation's Park, containing man and beast, in all the wild and freshness of their nature's beauty!'

But it was not until 1872 that Yellowstone Park in the American Rockies was established as the world's first National Park; a 'pleasuring ground for the benefit and enjoyment of the people'. Back in the old country, it was a Scottish Liberal MP then representing the distinctly urban constituency of Tower Hamlets who first raised the issue of National Parks. James Bryce had failed to win support for his Access to Mountains Bill of 1884, but in opposing the Ambleside Railway Bill of 1887 remarked:

> The people look upon the Lake District as their National Park, and they desire to preserve it. Are we to show ourselves less sensitive to the value of natural beauty than the people of California, who have set apart their Yosemite Valley to be kept sacred from the intrusion of railways? Are we to fall behind them by allowing works to be made which will destroy what Nature has bestowed?

ROBBING THE POOR OF THEIR PLAYGROUNDS

Bryce was supported by Henry Howorth MP, who wished to preserve 'the only area in England where Nature can be seen without modern vulgar embellishment'. Another MP, Robert Graham, noted some of the potential social benefits in creating an English National Park:

> I protest against robbing the poor of the nation of one of their playgrounds. It seems to me that, like the Yellowstone Park, this Ambleside district might be most advantageously bought by the nation, and preserved for the people of our great towns as a Democratic playground. In this country the rich have their enclosed parks, in which the noble owner can walk or commune with his gamekeeper beneath the shadow of the neat white-painted board – 'No Trespassers'; but his poorer neighbour can only ramble about his well-flavoured open slum, and saunter

PAGES 94–5
The combination of natural and man-made beauty in the North York Moors is a hallmark of England's National Parks.

IDEA 10 • NATIONAL PARKS

Wordsworth's Guide to the Lakes *was originally written to accompany the Rev. Joseph Wilkinson's sketches of the Lakeland landscape, first published as* Select Views in Cumberland, Westmoreland and Lancashire *in 1810 when it included this image of* Dunmail Raise on the Ambleside Road.

between that and the public-house, where refreshment is provided both for his body and soul – for I am even prepared to advance that the poor have souls.

Henry Labouchère MP then entered the debate, summing up the general opposition to any Wordsworthian notions that landscapes could ever belong to the nation in a withering assertion of the primacy of ownership: 'I protest against these attacks upon property. Is my Hon. Friend prepared to pay for the national playground he is going to establish for the benefit of sentimentalists, poets, and aesthetes?'

Despite generating support for the idea of English National Parks, Bryce persisted to alienate the House of Lords – who wished to protect their grouse and deer from wandering boots – by promoting his Access to Mountains Bill (for Scotland) on a yearly basis.

A NATIONAL RIGHT TO NATIONAL SCENERY

In 1908, fellow Liberal MP Charles Trevelyan – brother of the eminent historian G.M. Trevelyan, and husband of the future CPRE co-founder, Mary Trevelyan (*née* Bell) – reintroduced Bryce's Bill, but extended its scope to cover England and Wales. 'The Cumberland Lake District is a free paradise,' he claimed, reminding the House of Commons of the 'hundreds and thousands of men and women scattered over England to whom the freedom of these hills was one of the chief joys and benefits of their life'. Trevelyan contrasted the Lakes with the Peaks around Sheffield, which had been almost entirely closed off to the workers trying to escape the 'dingy, smoky, colourless city' by landowners lacking the 'traditional public spirit and unexclusiveness' of their Lakeland counterparts. Trevelyan believed it was indefensible 'to exclude a large and increasing number of men and women who were seeking health and recreation in the best and most beautiful wild places of the country on the ground that another set of people, much less numerous, but with more opportunities, wanted it for their exclusive recreation'.

The Times feared that an emphasis on access alone might divert 'public attention from the desirability of securing tracts of landscape as permanent national possessions', pointing out the potential of existing National Trust acquisitions in the Lake District to provide 'the nuclei of a "reserve" comparable in purpose and in beauty with the American Yellowstone Park'. Trevelyan ultimately failed to convince the Lords of the merits of access, and the Bill was not debated again until 1924, when

Piers Thompson failed to convince his colleagues that encouraging 'the people in our great towns to know the country and to love the countryside' would arrest the decline of patriotic sentiment symptomatic of a shell-shocked nation.

DEFINING THE POSSIBILITIES OF ENGLISH NATIONAL PARKS

Thompson's idea that landscape could 'lift up' the working classes seemed to be borne out by Lord Bledisloe's 1925 dispatches from America, in which he noted that their National Parks 'constituted a most perfect holiday resort for persons of all classes engaged normally in strenuous work'. On his return, Bledisloe argued that making the Forest of Dean a National Park would provide similar benefits to England's labourers and 'overworked brain workers'.

As Bledisloe's report gathered dust, the Belgian Congo and South Africa both established their first of several National Parks. The vast areas of 'virgin' land in America and Africa had allowed their Parks to be defined by the need to preserve wilderness, essentially by keeping people out. The English had been attempting to import the

The view over Hope Valley in Derbyshire's Peak District was the kind Ramsay MacDonald would have had in mind when urging the House of Commons to 'recognise the national right of the population to enjoy national scenery' in 1908.

same idea while – through the Access to Mountains Bills – expecting it to allow people to flood in. Sir Herbert Griffin later described the muddle of a time when some wanted English National Parks' 'main purpose to be recreational' while 'others wanted public access to be strictly limited and controlled'. Wider aspirations suggested they could be anything from hotbeds of scientific research 'dedicated to the protection of nature and flora and fauna', to living museums with 'specially built villages and camps, so that townsmen would be encouraged to see the farmer at work'.

By 1926, Patrick Abercrombie recognised that English National Parks must focus firstly on the urgent 'preservation of "wild country" with universal appeal for civilised man', in the hope that greater access might follow. His treatise on *The Preservation of Rural England* warned that the 'wildest country' often had the largest mineral deposits, meaning that 'public ownership will have to step in before the business exploiter gets thoroughly to work'. Working closely with Abercrombie and the other co-founders of the Council for the Preservation of Rural England, Griffin later confirmed CPRE's early aims for England's National Parks:

According to Patrick Abercrombie's 1929 evidence to MacDonald's National Parks Committee, Crown lands like the New Forest were deserving of protection as 'primeval patches of country'.

> They were to be regions of our finest natural landscape bought into full public service – preserved in their natural beauty, continued in their farming use, and kept or made accessible for open air recreation and public enjoyment, and particularly for cross-country walking.

CPRE's wishlist for National Parks had to contend with the fact that, unlike in America and Africa, England's finest natural landscapes were subject to claims of ownership, and were therefore vulnerable to those asserting their rights to hunt, mine or intensively farm this land. While foreign National Parks were large-scale and distinct landscapes, making it easy to define their boundaries and organise their administration, most potential English National Parks straddled several counties and were under the historic ownership of a variety of landed families. These landowners also just happened to be the ruling elite of the country, with little or no desire to allow the nationalisation of their land.

This presented another huge obstacle for the practical application of National Parks in England: how could they incorporate the working landscapes that made estates – and their villages and market towns – viable, both through farming and field sports. Back in 1908, George Younger MP had warned that 'a good or a bad grouse season meant all the difference between a comfortable and an impoverished winter for a great many people'. The importance to

'Natural beauty is the highest common denominator in the spiritual life of today.'

the ruling elite of deer stalking and grouse shooting was emphasised by Prime Minister Herbert Asquith, in his attempts to arrange the following Parliamentary year at the end of 1908: 'The 12th of August is a dead end. It is regarded almost as a violation of the etiquette of society to be in London and not on the grouse moors after that date.' The predilection of the aristocracy to serve – and be killed – as officers in the First World War combined with increasingly punitive Death Duties and Inheritance Tax to make the finances of the great estates even more precarious, and the survivors even less likely to sacrifice their income from sport.

Herbert Griffin and CPRE realised that in order to get widespread support for National Parks, they would have to appease the landowners and 'balance profitable land use with preservation'. They recognised the unifying potential of the fact that England's finest countryside was a democratic creation: the result of 'the diligence of hundreds of generations' of landowners *and* commoners who 'had impressed their seal on the landscape'. Unfortunately, the prospect of National Parks seemed increasingly distant in the early years of CPRE, when the cost of compensating landowners was likely to remain politically unpalatable until the Labour Party could gain a Commons majority. Griffin tried to shift the debate away from matters of economy, promoting English National Parks as 'cultural landscapes' – a celebration of the beauty that had been shaped by man and nature – unlike the wholly natural landscapes of the National Parks in other countries.

A ROLE FOR THE STATE?

Aware that he had a decidedly full in-tray, CPRE's August 1929 memorandum to the new Prime Minister, Ramsay MacDonald, simply suggested that if the Government decided to look into the matter of National Parks, they would make recommendations. Although the election of a Labour Government in the summer of 1929 might have been expected to reduce the influence of the landowning lobby, CPRE must still have been pleasantly surprised when MacDonald almost instantly set up the Addison Committee to inquire into the desirability and feasibility of National Parks for the 'preservation of natural characteristics and improvement of recreational facilities for the people'.

Sensing their opportunity, CPRE used its national conference in Manchester in November 1929 to instigate 'National Reserve Committees' in potential National Park areas. The Lake District National Reserve Committee was formed that very weekend (becoming the Friends of the Lake District in 1934 and going on to represent CPRE in Cumbria) under the guidance of Abercrombie and John Dower, a young architect who had 'married into' National Parks that summer after his wedding to Pauline, daughter of Charles and Mary Trevelyan.

In December, Abercrombie and the geographer Vaughan Cornish gave CPRE's evidence to the Addison Committee, both stressing the importance of National Parks for the recreation of the whole country, and especially those living in cities. Abercrombie listed twelve distinct areas that might be most suitable for National Park status, saying: 'Of the twelve areas given in the list, the High Peak and the South Downs would appear to have the first claim from the population point of view; from the point of view of national interest and intrinsic beauty the Lakes and Snowdonia, Exmoor and Dartmoor.' Cornish assessed the locations which gave most 'recreative value from the beauty of their outlook, the refreshing qualities of air and climate and the active pastimes of the countryside'. He singled out the potential of the South Downs to 'be of great value for the recreational instruction of the young', education

Abercrombie believed that Exmoor had a strong claim for National Park status 'from the point of view of national interest and intrinsic beauty'.

being 'an aspect of national parks which has received too little attention in England, but constitutes no small part of their claim on the national exchequer'. In his 1930 book on *National Parks and the Heritage of Scenery*, Cornish argued that by creating National Parks, the State 'would help the population's latent appreciation of beauty blossom', and repair some of the damage resulting from the 'present careless indifference of the town tripper in his charabanc'.

In 1931, Addison reported in favour of National Parks 'to safeguard areas of exceptional natural interest against disorderly development and speculation' and 'to improve the means of access for pedestrians to areas of natural beauty'. But, in the words of Herbert Griffin, his report 'was pigeon-holed because of a financial crisis', allowing MacDonald's new National Government to ignore CPRE's calls to set up a National Park Authority as recommended by the Committee. G.M. Trevelyan took up the cause at CPRE's October conference. '£100,000 a year for five years should be put at the disposal of the national park authority,' said Trevelyan, arguing that the sum 'was small compared with the money spent on London parks by the State for maintenance alone'. Trevelyan stormed on:

> It is an unfortunate tradition of the British State, and particularly the Treasury, that none of the taxpayer's money should be spent on preserving the beauty of the countryside and landscape. Since, however, the State is, for purposes of national defence,

Unique among National Parks around the world, England's are truly working landscapes. From the outset, CPRE's campaign had argued that places like Malhamdale in the Yorkshire Dales should continue to be farmed while being opened up for public enjoyment.

communication and forestry, rapidly destroying natural beauty, it is only fair that it should do something in return to preserve some of it.

With the economy and national security preoccupying governments, there was little progress on National Parks in the 1930s, despite CPRE convening a Westminster Central Hall conference at the end of 1935, which resolved to press for a State-funded 'central authority' on National Parks. Broaching the funding issue, Patrick Abercrombie ventured that 'some of the most beautiful country is almost the cheapest land', making it 'more liable to be bought up and spoiled. If the State could find £100,000 for a Codex [Greek Bible], it should be able to find a similar amount for these areas.' Clough Williams-Ellis quipped that if a CPRE deputation went to see the Prime Minister, they would hope to meet 'the Mr Baldwin who is the vice-president of the National Trust, rather than the Mr Baldwin who was apt to countenance the giving of large tracts to the Army for manoeuvres, bombing, and so on'.

PLACES OF INSPIRATION AND REFRESHMENT

Political resistance was summed up by leading civil servant Sir John Maud's statement in 1937 that although there was a strong case for National Parks, 'Housing, Milk for Mothers and a cancer scheme' were higher priorities. In the face of such pragmatism, campaigners had started to look beyond economic arguments as early as 1932, when Vaughan Cornish's *Scenery of England* argued that National Parks would provide 'that close touch of nature which is needful for the spiritual welfare of a nation'. The following year, Abercrombie's *Town and Country Planning* argued that young people in England's urban areas were 'increasingly thirsting for wild nature as a contrast to ugly towns'. The Friends of the Lake District began speaking in even loftier terms, describing National Parks as 'national playgrounds of body and spirit for the common man' and 'places of inspiration and refreshment for the poets and artists and philosophers of the future'.

In the context of widespread publicity for Lord Aberdare's National Fitness Movement, G.M. Trevelyan argued that National Parks should be considered part of 'any national health scheme'. Aberdare sent a message of support to a 1938 meeting of CPRE's Standing Committee on National Parks, at which Lord Horder argued that 'national fitness was not a question of physical jerks', but should cater for 'the whole man' through the preservation of the amenities of life: 'fresh air, beautiful sights and sounds, and a certain amount of quiet'.

At the same meeting, Trevelyan argued that while other European nations began to define themselves through political ideology, racial supremacy or military might, the landscape was central to our national identity:

> Without vision the people perish and without sight of the beauty of nature the spiritual power of the British people will be atrophied. Natural beauty is the highest common denominator in the spiritual life of today. Yet now that it is most consciously valued, it is being most rapidly destroyed upon this planet, and above all in this island. At present, man is destroying natural beauty apace in the ordinary course of business and economy. Therefore, unless he now will be at pains to make rules for the preservation of natural beauty, unless he consciously protects it at the partial expense of some of his other greedy activities, he will cut off his own spiritual supplies, and leave his descendants a helpless prey for ever to the base materialism of mean and vulgar sights.

It was the strongest possible rebuttal to the argument that we could not afford to create National Parks; Trevelyan was saying that, physically, emotionally and economically, we could not afford *not* to.

NATIONAL PARKS AT WAR: A NEW SENSE OF PURPOSE

Realising they had to step up a national propaganda campaign before war came, CPRE commissioned *Rural England: The Case for the Defence* to be shown in cinemas. The six-minute film was premiered to the press in London in February 1939, after which the sci-fi novelist John Gloag remarked that 'it shows how much there is to save in England and how well worth saving it is'. Shown in nearly a thousand cinemas around England during 1939, the film was chosen by the British Council to represent the best of England at that year's New York 'World's Fair'. The film asked cinema-goers:

> Are you going to allow such scenic loveliness to be despoiled because you won't fight to preserve it? Are you willing to give up your grand inheritance? Will you always be content with confined spaces? Or are you going to take the road to National Parks? The road to freedom; freedom of England's country. That is the case for the defence of our right to the beauty of our land.

The film's language of fighting for freedom became hugely poignant with Britain's declaration of war on Nazi Germany in September 1939, and helped make National Parks resonate with politicians for the first time. Lord Reith, the wartime Minister of Works and future head of the BBC, incorporated 'the reservation of National Parks' in a national plan for reconstruction approved by Churchill's cabinet on the last day of 1940. Churchill then appointed Lord Justice Scott, a member of CPRE's Executive Committee, to look into 'Land Utilisation in Rural Areas'. Scott's 1942 report concluded that National Parks were 'long overdue' and must 'be preserved for the enjoyment of the whole nation'. Scott recommended that National Park boundaries should be drawn up within the first year of the peace, and in 1943, Planning Minister William Morrison gave the job of surveying potential locations to John Dower.

While conceding that his idea bore little resemblance to the American idea of National Parks as vast, untamed wildernesses, Dower's 1945 report gave the first official definition of an English National Park:

> An extensive area of beautiful and relatively wild country in which, for the nation's benefit, the characteristic landscape beauty is strictly preserved; access and facilities for public open-air enjoyment are amply provided; wildlife and buildings and places of architectural and historic interest are suitably protected, while established farming use is effectively maintained.

Addressing that most thorny issue, Dower added that 'the system of landownership will nowhere be allowed to stand in the way of a democratically determined allocation' of National Parks. In April 1946, the Labour Chancellor Hugh Dalton recalled the old Liberal song 'The Land' when he announced the creation of a National Land Fund, at which point Churchill offered to give the Commons a rendition of the tune. The National Land Fund comprised £50 million 'to increase the National Estate', including the 'creation of National Parks'. Reflecting that much of the country had been 'spoiled and ruined beyond repair', the country-loving Chancellor – known as the 'red rambler' – said that which remained, 'the deep peace of the woodlands, the white, unconquerable cliffs . . . should surely become the heritage, not of a few private owners, but of all our people'.

Despite overwhelming political and public support it began to look as if there might not be time for National Parks legislation before the 1950 election. The huge changes the Government were attempting – including the nationalisation of industry, the creation of the NHS, the building of New Towns and the reconstruction of old ones – were seen as more urgent. The tragic and untimely death of John Dower from tuberculosis was another huge blow, and may have explained the stalled progress; his fellow campaigner Ethel Haythornthwaite said that when Dower died 'the main inspiration [for National Parks] in the Government died too'.

With a near twenty-year campaign threatening to peter out in the most heart-breaking fashion, Tom Stephenson – the Ramblers' Association Secretary – organised a three-day walking trip along the Pennine Way in 1948 for the prominent Labour Party ramblers. The trek, and the ensuing publicity, evidently had the desired effect of stiffening Hugh Dalton's resolve, and the National Parks and Access to the Countryside Bill was finally introduced in March 1949.

A PEOPLE'S CHARTER FOR THE OPEN AIR

As Lewis Silkin proudly announced in introducing an unopposed second reading, this was an extraordinary piece of legislation: 'Now at last we shall be able to see that the Lakes, the moors and dales of the Peak and the tors of the West Country belong to the people as a right and not as a concession. This is not just a Bill. It is a people's charter – a people's charter for the open air.' The National Parks and Access to the Countryside Act received Royal Assent on 16 December 1949, establishing the National Parks Commission to designate 'extensive tracts of country' offering 'natural beauty and opportunities for open air recreation' as National Parks, and smaller areas of comparable merit as Areas of Outstanding Natural Beauty.

Herbert Griffin was appointed a National Parks Commissioner, and worked closely with the Commission's first Secretary, Harold Abrahams. This was the very same Harold Abrahams who won the 1924 Olympic 100-metre final for Britain, as chronicled by the film *Chariots of Fire*, produced by future CPRE President David Puttnam. During the Commission's first decade, Griffin and Abrahams helped designate seven National Parks in England and three in Wales, all of which reflected the locations specified by Patrick Abercrombie in 1929. The first was the Peak District, established in 1951 and now the second most visited National Park in the world after Mount Fuji in Japan. In 1957, Abrahams received a CBE for his contribution, while Griffin – already a CBE – received a Knighthood in the same honours list.

Just as Abercrombie had prioritised preservation over access, Abrahams measured the success of National Parks in terms of what they had preserved, as editor of *Britain's National Parks* in 1959:

> The visitor cannot see the damage which has been prevented. He will not see the blanketing of a wide landscape in the North York Moors by a large conifer plantation; he will not see any overhead lines in Lower Borrowdale, one of the loveliest Lake District Valleys; nor will the visitor see at Lee Moor, Dartmoor, china clay workings eating away a conspicuous moorland ridge, nor the hundreds of unsuitable smaller developments for which Park Planning Authorities have refused permission.

Pauline Dower closed Abrahams' book by reflecting on the positive benefits that preservation – and the work of her late husband John Dower – had made possible. She believed that National Parks had given the people of England a 'sense of belonging, of having roots in our country's past', and that life without them would be a 'meaningless existence'.

The Norfolk Broads became England's eighth National Park in 1989 with the New Forest following in 2004; the South Downs – an eighty-year CPRE campaign given a final push by their then President Bill Bryson –

IDEA 10 • NATIONAL PARKS

The South Downs finally gained National Park status in 2009, after eighty years of campaigning by CPRE.

finally completing Abercrombie's original list in 2009. National Parks remain the landscape jewels in England's crown, giving the highest possible level of protection for nearly 10 per cent of the country. And yet, in 2011, the Government proposed that they should promote more development, despite research showing that, far from stifling development, they are already exemplars of thriving rural economies: England's National Parks generated £10.4 billion for the Exchequer in 2012, supporting 68,000 jobs in tourism and 17,300 in farming.

National Parks receive over 90 million visitors a year, while a 2012 survey found that 93 per cent of the public view them as areas of national importance. And they are increasingly becoming important in ways that their historic advocates and administrators could never have conceived of: the soil in our National Parks stores over 119 megatons of carbon – equivalent to England's entire annual CO_2 emissions.

That these pioneers managed to create ten unique, beautiful and much-loved National Parks in England's crowded island is an incredible achievement. They are now, truly, the sort of national property Wordsworth of which would heartily approve.

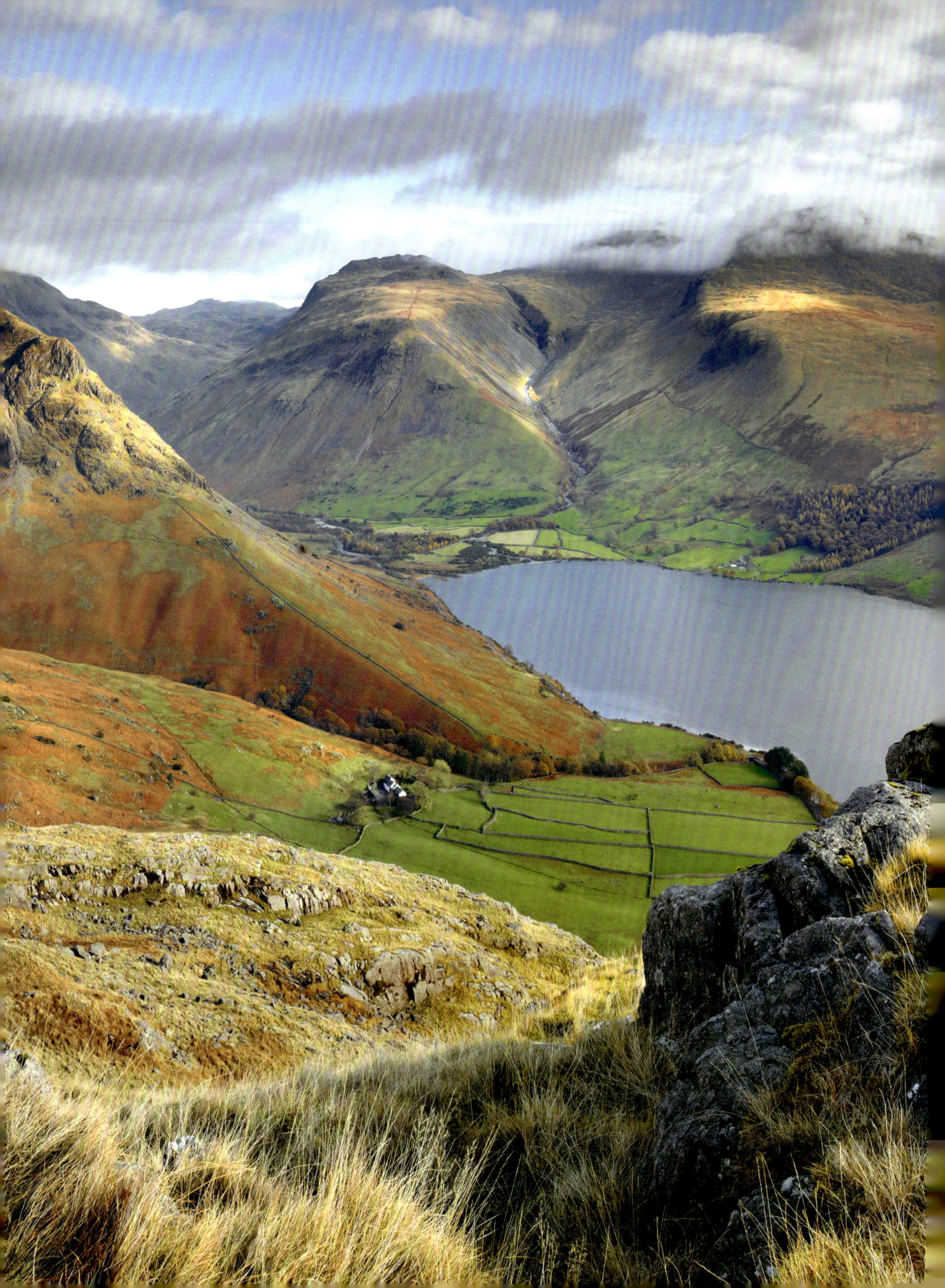

Idea 11

The Sense Sublime

introduced by Andrew Motion

Inside us all, wherever we live, however familiar we might or might not be with the countryside, is an absolutely primal atavistic need for green places and open spaces. The more we are bombarded by the demands of modern life, the more important it becomes to enjoy peace and quiet: more darkness, more solitude, more beauty, the pleasures of uncluttered ground. In the countryside time slows down, longer perspectives open, richer thoughts accumulate – because in the countryside we enjoy the essential things about being human.

The language of Wordsworth's 'Tintern Abbey' feels very familiar to anyone who has ever grappled with that sort of pulsing sense of connection that we feel standing in a beautiful field or watching a beautiful sunset. We think that we need to reach for very large concepts, like 'The Sublime', in order adequately explain the feelings that we're having in those moments, but Wordsworth's 'sense sublime' can be common to all of us. To everyone who looks out of the train window and gets a shiver of pleasure, to the poets in their attics, to the city-dwellers in their parks, to the farmers in their fields. To everyone.

Wordsworth showed how looking very hard at particular objects – particular stones, particular leaves, particular trees, particular blades of grass – can allow us to see into the life of things, to feel a connection with the earth that is as deep as the deepest love we feel for another human being. Our sense of who we are as English people, whether we live in the countryside or not, depends in a very fundamental way on this connection with the bedrock.

It may have been a Swiss philosopher, Jean-Jacques Rousseau, who ushered in the age of the Romantic poets and artists who placed their faith in their feelings and revered the mysteries of the natural world. But it was in England where Nature and her secrets came to be seen as something with huge spiritual and emotional significance; the English countryside became a place to commune with our innermost thoughts and our own personal connections to the land. If we had never learned to use the landscape's beauty and tranquillity to tap into Rousseau's 'happy state of ignorance' and form an emotional bond with the countryside, the case of the campaigners trying to protect it would have been much less compelling.

The Irish philosopher Edmund Burke distanced himself from Rousseau and Romanticism, but his contention that our judgment of beauty required only the instant response of 'the senses and imagination' seemed to agree that the intellect was no substitute for feeling. This idea was incredibly influential on a fledgling Romantic movement which placed its faith in emotion, rather than science, to establish the truth.

Burke's *Philosophical Enquiry* of 1757 introduced his concept of the Sublime as the state of astonishment brought on by the suspension of reason in the face of the immensity of Nature. Long before Burke had given the phenomenon a name, in 1712, the poet and politician Joseph Addison had been one of the first to describe the pleasure of being becalmed by 'stupendous works of nature'; writing in the magazine he had founded, *The Spectator*, Addison remarked that 'our imagination loves to be filled with an object, or to grasp at anything that is too big for its capacity. We are flung into a pleasing astonishment at such unbounded views, and feel a delightful stillness and amazement in the soul at the apprehension of them'.

In 1778's *Reveries of a Solitary Walker*, Rousseau gave a real-life example of the power of the Sublime, and the distress that could be caused if this transcendental state of mind was interrupted. Enveloped in the trees and birdsong of the Alps, Rousseau became distracted by a faint but repetitive mechanical clicking, before discovering a stocking factory just 'twenty yards from the very place where I had thought to be the first person to tread'. The *Reveries* emerged posthumously, becoming a huge influence on the English Romantics hoping to experience Rousseau's blissful loss of self-consciousness by drinking 'deeply of the beauty of nature' and its 'fresh shady spots, brooks, thickets, and greenery'.

WORDSWORTH'S DEEPER CONNECTIONS
William Wordsworth was just eight years old when Rousseau died in 1778 and has been described as a spiritual heir of the philosopher. In *The Table's Turned* (1798), he joins Rousseau and Burke in rejecting the objectivity of reason for the subjective feelings of the heart:

PAGES 108–9
Sublime landscapes like Scafell Pike in the Lake District stir our deepest emotions and allow us to connect to the earth.

IDEA 11 • THE SENSE SUBLIME

The Langstrath Valley typifies the sublime power of the Lake District which fired the imaginations of the English Romantics and continues to inspire us today.

Enough of Science and of Art;
Close up those barren leaves;
Come forth, and bring with you a heart
That watches and receives.

But where Nature purified Rousseau's imagination, for Wordsworth, Nature – and all its minute details – *was* the imagination; in his beloved Lake District, inspiration was as likely to appear – like an 'unfather'd vapour' – at the site of a humble daffodil as a great mountain. Wordsworth's poems did not just observe the landscape, they allowed him to communicate directly with it, or even become part of it in the conviction that Man and Nature were inseparable. His great theme was the connection between the physical landscape and the territories of the mind and heart, a connection that produced a depth of meaning which could not be defined or explained by science.

In his 'Lines Composed a Few Miles above Tintern Abbey' of 1798, he speaks of feeling, in Nature:

> A presence that disturbs me with the joy
> Of elevated thoughts; a sense sublime
> Of something far more deeply interfused

For Wordsworth, landscape provided a chance to connect with an original, tranquil state of being – whether real, imagined or remembered. He famously defined poetry as 'the spontaneous overflow of powerful feelings: it takes its origin from emotion recollected in tranquillity'. Using the landscape as a portal to access these recollections, 'Tintern Abbey' combines the immediacy of rediscovery with the warm memories of an earlier visit with his sister Dorothy, letting the

> . . . wild secluded scene impress
> Thoughts of more deep seclusion; and connect
> The landscape with the quiet of the sky

The paintings of Wordsworth's close contemporary John Constable used the landscape in a similar way. After moving to London in 1816, he found he could transport himself back to childhood for scenes like 1821's *Landscape: Noon* – more commonly known as *The Hay Wain*. In 2003, Lucian Freud praised Constable's ability to mine his childhood haunts, saying this intimacy gave Constable a 'depth of possibility that has more potential than seeing new sights'. Constable asserted that 'painting with me is but another word for feeling'; indeed, his later admirers *felt* these pictures in the same way that Constable felt the source material. Kenneth Clark later argued that Constable's use of 'nature as the material means through which pictorial emotion can be expressed' helped subsequent generations discover 'our own emotions and learn to re-create them as shape and colour'. While Thomas Hardy recreated his emotions for the landscape in words, 1874's *Far From the Madding Crowd* showed that he shared Constable's – and Wordsworth's – understanding that a 'spot may have beauty, grandeur, salubrity, convenience; but if it lacks memories it will ultimately pall upon him who settles there'. Hardy later asserted that 'it is better for a writer to know a little bit of the world remarkably well than to know a great part of the world remarkably little'.

THE SPIRITUAL HOPE OF BEAUTY

By the 1840s, Wordsworth's creative force was most likely to be roused by the prospect of a new railway being built through his beloved Lakes – often to satisfy the demand for tourism he had unwittingly helped to create. In *Furness Abbey* of 1845, he notes that the 'simple-hearted' navvies building the new line to Barrow were able to 'feel the spirit of the

When Constable painted Stonehenge in 1835, he wrote: 'it carries you back beyond all historical records into the obscurity of a totally unknown period'. CPRE's founding President Lord Crawford later campaigned to save the 'lonely majesty' which allowed this sense of connection with the past.

place', while the 'profane despoilers' – the 'civilised' railway investors – remained unmoved. In the following year's *Modern Painters*, John Ruskin added that not only artists and poets, but even 'those of the least thoughtful disposition' were capable of experiencing the 'deeper feeling of the beautiful' and the 'spiritual hope and longing' Nature could inspire.

Providing a link between Ruskin and the later converts to rural spirituality was someone who spent the first twenty-eight years of his life in Argentina. The formative experiences of the flat, grassy Pampas around Buenos Aires gave William Henry Hudson a unique perspective on the English countryside. On a fundamental level, rural landscapes became a respite from the 'strangeness' of the urban society he encountered on arriving, without a penny to his name, in the lower echelons of Victorian London. Soon becoming a successful rural writer with a gift for articulating the importance of wild birds to the countryside, Hudson became a founding member of the RSPB in 1889. In 1903's *Hampshire Days*, he describes how he learned to embrace the land on a deeply spiritual level: 'my flesh and the soil are one, and the heat in my blood and in the sunshine are one, and the winds and the tempests and my passions are one.' *A Shepherd's Life* (1910) highlighted the depth of this spiritual connection among those who followed in the footsteps of generations of ancestors who 'flourished and died like trees in the same place'.

By the 1920s, such ideas of rural spirituality were second nature to the likes of Patrick Abercrombie and Lord Crawford. Abercrombie's *Preservation of Rural England* looked further east for fresh inspiration, noting the Chinese 'intense veneration for Natural scenery' and their belief that 'man is merely a temporarily detached and animated fragment of the earth'. In 1927, CPRE President Lord Crawford (acting in his capacity as President of the Society of Antiquities) appealed to save the solitude of Stonehenge's landscape setting, so that 'our posterity will see it against the sky in the lonely majesty before which our ancestors have stood in awe throughout all our recorded history'.

For Vaughan Cornish, his work for CPRE in campaigning to preserve the countryside was 'not merely for the pleasures of the eye, but for the deepest satisfactions of the soul'. He believed Wordsworth had showed us how 'wandering in Nature's solitudes' could help overcome the belief – ingrained by centuries of civilised doctrine – that the countryside existed to be conquered: this spiritual revelation could be brought on simply by

The countryside was 'not merely for the pleasures of the eye, but for the deepest satisfactions of the soul'.

experiencing the landscape as a place where 'the tumult of busy thought die down and the faculties of direct perception obtain their opportunity'.

G.M. Trevelyan's 1931 lecture on the *Call and Claims of Natural Beauty* argued that this direct perception allowed us to feel 'subtle and strange emotions' prompted, for instance, by 'the eternal recurrence of spring' which 'fills us with a sense of joy, more primeval and powerful than the mere delight' in beauty. By 'shutting himself up in cities' Trevelyan felt modern man was denying himself the 'sights and sounds that are natural to him by an infinitely long inheritance'; in other words, we had evolved to love nature instinctively. It is this instinct that means even today plans to sell-off ancient forests or historic landscapes generate a national outcry and a real sense of personal loss, even among urbanites who have not set foot in the country for years. Trevelyan argued that natural beauty could claim to be above religion and science as 'the highest common denominator in the spiritual life of today', partly because religion and science had created so much uncertainty in each other: 'man looks round for some other source of spiritual emotion that will not be either a dogma or a fashion . . . something far older yet more fresh, fresh as when the shepherd on the plains of Shinar first noticed the stern beauty of the patient stars.'

THE MODERN ROMANTICS

For a brief time, in the run up to the First World War, English artists and poets had been tempted, by the fashions of Futurism, to re-seal the portal Wordsworth had established in the landscape. C.R.W. Nevinson was a leader of the new movement, but in the wake of the war even he called for a 'reactionary, to save contemporary art from abstraction'. In the 1920s, two of the first to respond to the emotional tug of old, rural England were John Betjeman and John Piper, who were promptly accused by their contemporaries of betraying the modern movement. In a 1935 radio debate, the architect Sir Reginald Blomfield predicted the imminent demise of crude experimentation followed by the restoration of art as the 'expression of the permanent instincts and emotions of man'.

Blomfield's 'restoration' would be influenced by W.G. Constable's catalogue notes of the Royal Academy's 1934 *Exhibition of British Art*, which argued that English painters painted in response to the muse of 'the physical and emotional call of a particular place or person'. When

John Constable painted familiar places to access childhood memories, he could not have imagined that a distant relative would later describe this work as a 'humanised concomitant to the affairs of everyday life'. The same year as the Royal Academy exhibition saw Paul Nash place himself firmly in the tradition identified by W.G. Constable; his discovery of Avebury's standing stones encouraged him to seek further connections to the ancient spirit of the landscape by 'hunting far afield over the wild country to get my living out of the land as much as my ancestors ever had done'.

Nash was better known as the founding member of the progressive Unit One collective which, in his letter to *The Times* of 1933, rejected English art's 'Nature cult' and lack of 'structural purpose'; this was the same Nash who painted a relatively conventional watercolour of the Wittenham Clumps – an ancient British hill camp in the Berkshire Downs – just two years later. Nash had discovered the Clumps – and with them his love of landscape – aged nineteen, returning throughout his life to paint them with a realism far removed from his shockingly modern work as a war artist. Regardless of the method of painting, Nash believed the spirit of the land was the defining characteristic of English art. He devoted the last year of his life to a series exploring his feelings towards the mysteries of Nature through the Clumps whose importance he said he *felt* 'long before I knew their history. They were the pyramids of my small world.'

By 1944, the Second World War had encouraged people to make links between England's threatened landscapes and its greatest artists, poets and musicians. Alexandra Harris describes how Bill Brandt's 1940s photographs of 'literary landscapes' began to sanctify parts of the countryside to the point that 'visions of Piers Plowman start to appear' on entering William Langland's Malvern Hills. Similarly, Yehudi Menuhin once confessed he was unable to walk the Malverns without feeling the presence of Edward Elgar. The landscape was so integral to the composition of his *Cello Concerto* that Elgar had once said that if its melodies ever seemed to float across the hills, 'don't be frightened – it's only me'. When conducting a rehearsal of a gentler passage of his *Symphony No. 1*, he asked the orchestra to 'play it like something you hear by the river'. Naturally, Elgar became the Honorary Vice President of CPRE's Herefordshire branch from July 1931 until his death in 1934.

Elgar's contemporary Frederick Delius died shortly after Elgar, but not before confirming the landscape connections of his music, saying 'much of my inspiration has come from the moors. I hope to get one more whiff of them before I die.' The third great English composer to die in 1934, Gustav Holst, had been inspired to write a symphony for the Cotswold Hills he once listed as a reason to feel grateful to be alive, alongside music and his walking partner, Ralph Vaughan Williams. Despite a 'craven spirit' he supposed 'comes of being born in Surrey', Vaughan Williams confessed that 'my heart goes through the same manoeuvres as Wordsworth's when he saw a rainbow when I see a low range of hills'. Living in London at the turn of the century Vaughan Williams took to collecting old English folk songs, attracted by the prospect of spending time in country inns 'in the rare company of minds imbued with that fine sense which comes from a life-long communion with nature'.

RURAL ROOTS

If Menuhin suspected Elgar's tunes were written by the hills, he also said that 'if wind and water could write music' it would sound like Benjamin Britten. Britten returned to England from his new life in California after reading an E.M. Forster article on how George Crabbe's poetry had been inspired by the Suffolk countryside. Crabbe's poetry transported Britten across the Atlantic, and 130 years back in time; the everyday details of Crabbe's landscapes created a literary portal through which Britten was drawn back to what he called 'the realities of that grim and exciting sea-coast'.

On becoming a Freeman of Lowestoft in 1951, Britten ended his acceptance speech with lines which illustrate why our sense of rootedness

Worcestershire's Malvern Hills: a portal to the *Visions of Piers Plowman* and the ghostly sounds of Elgar's *Cello Concerto*.

to the landscape has made the defence of the countryside a national passion: 'I treasure these roots, my Suffolk roots; roots are especially valuable nowadays when so much we love is disappearing or being threatened, when there is so little to cling to.'

In 1970, the future Poet Laureate Ted Hughes declared that a 'monstrous . . . nearly-autonomous Technosphere' was the result of the Western mind's exile from Nature, creating a world where the most common connection to the land came through 'the Developer peering at the field through a visor'. The following year, Philip Larkin's *Going Going* mourned:

> The sense that, beyond the town,
> There would always be fields and farms,
> Where the village louts could climb
> Such trees as were not cut down;

With the pace of change of modern life and all its uncertainties, it is not surprising that so many of us share Larkin's fear that, soon, our own personal bits of England will be gone, and recognise the sense of spiritual displacement Britten felt while living in California. And most people familiar with urban living will have witnessed, or experienced, the 'follies, madnesses and meanness' Trevelyan attributed to those 'shut up in cities'. The landscape gives us a shared cultural heritage, rooting us in the natural world and calming us with its sheer timelessness. The philosopher Anthony O'Hear has recently echoed Trevelyan's argument that humans have only recently adapted to urban life, contending that we still feel, on a primal level, 'a sense of physical exhilaration in emerging from a town into the open countryside'.

England's landscapes are tapestries of our total acquired human understanding and experience. We all came from the land, and we all have an instinctive desire to go back there. We may not be able to connect with the level of empathy shown by the rural poet John Clare – who could articulate the primitive emotions of the birds and animals around him, or sense the pain of an 'injured brook' – but we can all follow Wordsworth in engaging not just our eyes, but our whole selves, with the landscape. That's why countryside with deep cultural connections is the most stoutly defended of all, and why the case to make the Lake District a World Heritage Site made so much of the area's links to Wordsworth. It is how we came to understand that the countryside is vitally important as the most democratic and accessible way to experience the highest forms of enjoyment, inspiration and tranquillity. As Andrew Motion said in his first speech as CPRE President: 'Wordsworth's "sense sublime" can be common to all of us, wherever we've been brought up, and however long our families have lived where they live now. It is the sense that should be available to every walker and hiker and camper.'

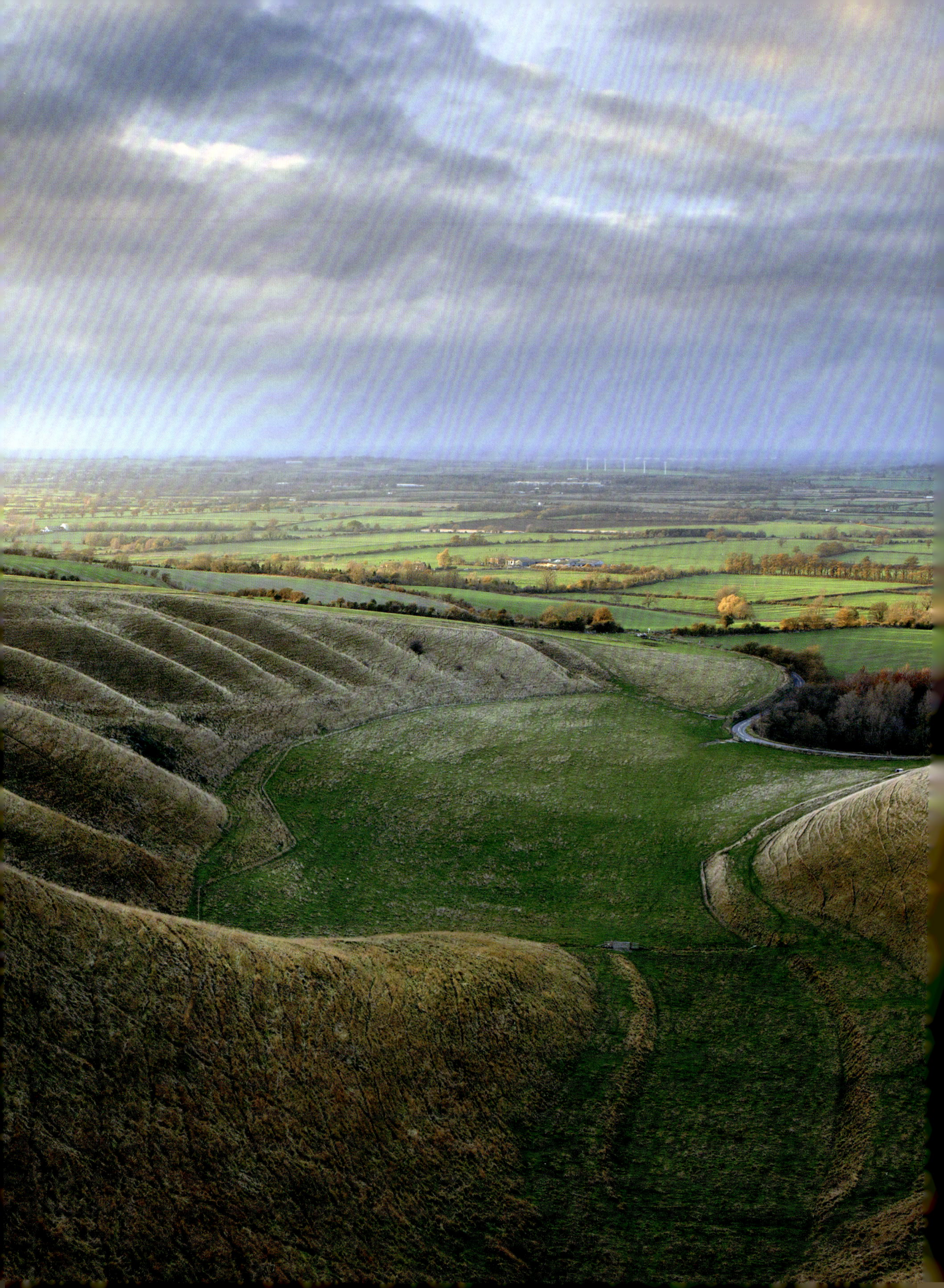

Idea 12

The Green and Pleasant Land

introduced by Clive Aslet

Yes, the rural idyll is infinitely seductive to us. As the authors show, there is – or has been historically – a mismatch between an idealised vision of the countryside, as expressed, for example, in the Arcadia created by eighteenth-century landscape gardeners, and the misery and remorselessness of agricultural labour, poignantly revealed in the works of the unhappy 'Thresher' poet Stephen Duck. But certainly our imaginative and emotional response to the countryside has sharpened our passion to preserve it.

Has the dichotomy now been spanned? For the man in the air-conditioned, stereo-equipped, computer-driven cab of the combine harvester, work isn't the round of drudgery that it was, even in the mid-twentieth century. And many people who live in the countryside have nothing to do with it as a place to produce food. They can live the idyll . . . provided, of course, that their view isn't blocked by others who have come to do the same. What a conundrum it has created!

From Shakespeare's 'other Eden' to Stanley Baldwin's 1924 assertion that 'England is the country and the country is England', centuries of political propaganda and poetic licence have created a sense of nostalgia for a rural idyll, which may never have existed quite as we imagine. Whatever the realities of our rural past, this cultural imprinting has made green fields part of our psyche, while ugly and unimaginative development instinctively offends our senses. While the London 2012 Olympics were, on the face of it, a celebration of Britishness, it was the portrayal of England's countryside during the opening ceremony that seemed to capture the public's imagination most. Several major opinion polls that year suggested that the countryside was as important to our national identity as the monarchy, Shakespeare and the NHS; as Simon Jenkins put it, 'the English countryside is embedded in the national personality.' And yet 80 per cent of us live in urban England and, as the Prince of Wales remarked in 2014, 'Many people in the UK are now four or more generations removed from anyone who actually worked on the land.' The Prince conveyed a glaring disconnect, which we all recognise – that even those with only a vague understanding of rural life 'treasure the countryside . . . we value so highly'.

The foundations of the rural idyll are found in Elizabethan attempts, in the words of the philosopher Roger Scruton, to 're-enchant the land'. The Norman Conquest had fundamentally changed the land: ownership and rights had become more important than common freedoms, while ancient rural customs were criminalised overnight, leading to bitter disputes, harsh punishments and generations of rural poverty. After five centuries of conflict, the Elizabethan Age allowed the country to be reborn in a state of peace, harmony and beauty. The countryside was the most obvious vehicle for these ideals; Scruton suggests that by taking refuge in the pastoral and the idyllic, the Elizabethan renaissance aimed to 'internalise the topography of England, to remystify it and to deliver it up as a home'.

The Elizabethans created the myth of the English patchwork as a unifying national emblem, to which everyone could belong. Shakespeare helped promote this narrative in John of Gaunt's dying lament in 1595's *Richard II*, conveying the idea that if only its rulers could be less obsessed with personal gain (the 'inky blots and parchment bonds' of rents and taxes), England's 'other Eden, demi-paradise' – the countryside – could be a unifying force.

ENGLAND'S GOLDEN AGE

Five years earlier, Sir Philip Sidney's *Arcadia* recreated the imaginary paradises of the classical 'Golden Age'. Virgil's references to the beauty of Britain in his *Eclogues* prove that rural England was being idealised as early as 38 BC for its 'mossy springs, and grass more soft than sleep'. Virgil was a huge influence on *Arcadia's* 'refreshing silver rivers' and

PAGES 118–19
The idea of 'the green and pleasant land' is almost as old as the prehistoric landscape of 'The Manger' at Uffington, Oxfordshire.

OPPOSITE
The ancient Arcadian topography of Cranborne Chase, on the border of Dorset and Wiltshire.

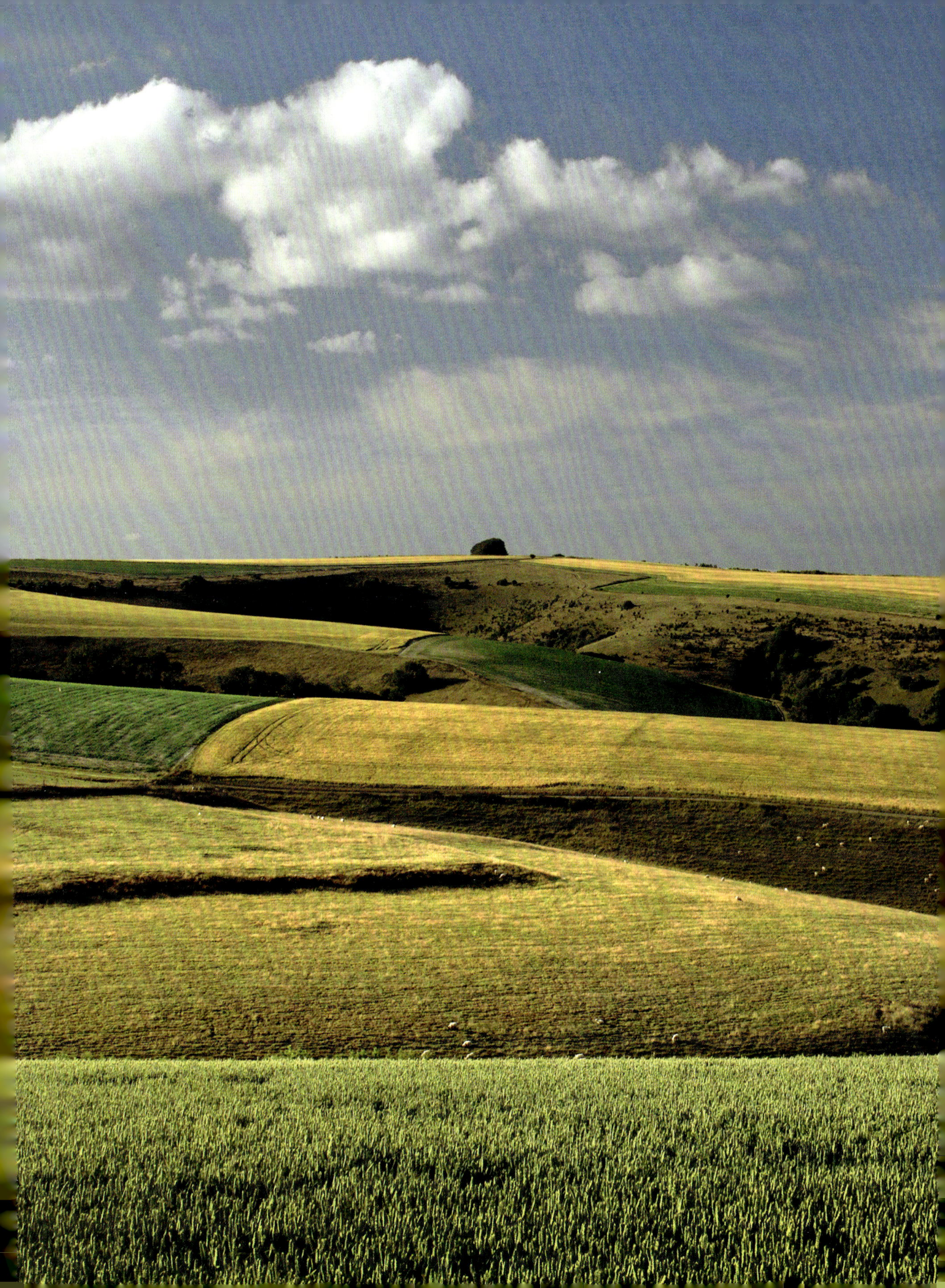

'cheerful, well-tuned birds', as was John Leland's early-Elizabethan description of England's 'little shady groves and meadows strewn with flowers'. Although Sidney's story was, on the face of it, an ancient Greek drama, its full title – *The Countess of Pembroke's Arcadia* – reveals its genesis in the Wiltshire countryside.

The titular Countess happened to be Sidney's sister, and the Arcadia was – in the words of the antiquarian John Aubrey – a celebration of her estate's 'romantic plains and [copses]'. The sense that this landscape represented a perfect vision of England, 'a heavenly dwelling', percolated through all levels of society, with lines from the book even quoted by Charles I on the scaffold in 1649. Aubrey would describe Sidney's real life Arcadia as 'a Paradise', where 'the innocent lives of the shepherds do give us a resemblance of the golden age'. But Sidney admitted that even the perfection of this original English idyll was a poetic creation: 'Nature never set forth the earth in so rich tapestry as diverse poets have done . . . her world is brazen, the poets only deliver a golden.'

By 1675, Izaak Walton was daring to question whether the true England was less a 'garden of piety, of pleasure, of peace' than the 'thorny wilderness of a busy world'. But the countryside itself remained a virtuous ideal, which could not be corrupted by the evils of the city; this convenient narrative proved a rich seam of inspiration for so much eighteenth-century art and literature, beginning with Hogarth's iconic *Rake's Progress* of 1732, which portrayed rural innocence lost to the temptations of urban life. William Cowper famously suggested that 'God made the country, and man made the town' in 1784; two decades later William Blake's introduction to his poem *Milton* imbued the English countryside with even greater holiness, by perpetuating the idea that Christ had been 'on England's pleasant pastures seen'.

Blake's idea of contrasting of London's 'dark satanic mills' with the purity of the rural landscape gained an incredible hold on the English sense of self. A lifelong Londoner, Blake's relocation to the village of Felpham between 1800 and 1803 allowed him to experience the English rural idyll for himself. 'Away to sweet Felpham, for Heaven is there,' he wrote back to London. 'The sweet air and the voices of winds, trees, and birds, and the odours of the happy ground, make it a dwelling for immortals.' Despite the impact of Blake's 'green and pleasant land' there was growing awareness among the new generation of English social reformers of the less than idyllic reality of rural life, even among the most beautiful surroundings. During his *Rural Rides* of 1826, William Cobbett was captivated by Sidney's Arcadian Wiltshire, where even on a wet winter's day 'the valleys are snugness itself'. However, the extreme poverty of the local workers soon became apparent, causing Cobbett to remark that 'in taking my leave of this beautiful vale, I have to express my deep shame, as an Englishman'.

IDEA 12 • THE GREEN AND PLEASANT LAND

Wenlock Edge in the Shropshire Hills became synonymous with the 'Golden Age' of nature writing heralded by A.E. Housman.

OUR COMMON HERITAGE

The late-Victorian period was one of enormous enthusiasm for England's countryside as the birthplace of two of the strongest facets of our shared national culture. The commons, though rapidly disappearing, had always allowed the English to feel they had a stake in the countryside, while mysterious antiquities like Stonehenge exposed the land-worship of our ancient forebears. Both commons and antiquities were being hotly debated in the press and in Parliament from the 1860s, helping to shape the idea that the countryside was central to our national identity. For the ancient monuments campaigner and CPRE founding President Lord Crawford they reminded modern society that 'primitive man was dominated by his environment and the yield of its soil', and that every English landscape had been 'impressed with the seal of laborious activity' of our long-vanished ancestors.

In 1893, Kenneth Grahame captured this feeling in his *Pagan Papers*, published fifteen years before *Wind in the Willows* and thirty-three years before he became a founding member of CPRE. He felt that modern England co-existed with 'the old England of heath and common' and an ancient England that seemed almost tangible 'on the downs where Alfred

These day trippers would have been deeply influenced by the 'enormous gush of country sentiment' the nation felt in the 1920s.

fought . . . gazing up at the quiet stars that had shone on many a Dane lying stark and still a thousand years ago'. For Grahame, the ancient Ridgeway of the North Berkshire Downs told the 'quiet story of the country-side' while 'the hurrying feet of the dead raise a ghostly dust'.

The formation of the Commons Preservation Society in 1865 and the passing of the Ancient Monuments Protection Act of 1882 bookended two decades that helped reclaim and celebrate the best of England's rural heritage, and bought it into sharp relief with the worst consequences of industrialisation. For the late-Victorian artists and preservationists, according to historian David Cannadine, 'the English countryside was made idyllic because in contrast to the squalor and deprivation of the towns, it was the very embodiment of decency, Englishness, national character and national identity.' The determination to venerate and preserve our shared past has been a national trait ever since.

The new Georgian era seemed to herald another Golden Age of nature writing, with the 'Dymock poets' of Gloucestershire – including Robert Frost, Edward Thomas, and Lascelles Abercrombie (brother of CPRE founder Patrick) – complementing the sentiments of Brooke, Belloc, Hardy and Housman. In 1912, Price Collier gave an analysis of *England and the*

English from an American Point of View, venturing that it was not London, Parliament or the Empire that was most precious to an Englishman, but its 'green fields, hills, valleys and hedges'. Anyone who dipped into England's literary output of the time would have agreed with Collier's judgement that 'this soft landscape is the England men love'.

After the war, George Orwell described how the nation had become urban enough to forget the hardships of rural life, but had remained close enough to respond to 'an enormous gush of country sentiment'. Knowing *A Shropshire Lad* by heart as a seventeen-year-old in 1920, Orwell understood why the 'idealised rustic . . . the charm of buried villages' and 'nostalgia of place names' had a deep appeal until the late 1920s. For Orwell and other middle-class boys of his generation, the 'picturesque side of farm life' was brought to life through the poetry of ploughing, fuelling the fantasy of a more 'primitive and passionate' life of 'rabbit-snaring, cock-fighting, horses, beer and women'. Of course, Orwell admitted that this romantic interpretation was only possible because he had not noticed the reality of rural life – 'the horrible drudgery of hoeing turnips and milking cows with chapped teats at four o'clock in the morning'.

For Vita Sackville-West, in her epic 1926 poem, *The Land*, one did not have to endure the drudgery to feel part of it. She understood that even those who had never heard 'the sheep-bells in the fold' could still succumb to the rural idyll:

> The Country habit has me by the heart,
> For he's bewitched forever who has seen,
> Not with his eyes but with his vision . . .

In her later introduction to *Kent Today*, she concedes that the mention of the Garden of England 'evokes most readily . . . acres of blossoming trees, short sunny slopes of chalk hills and wide skies and lush meadows', but asks 'how true, in actual fact, is this idyllic picture?' Sackville-West admitted 'that I have dwelt on the favoured corners and have left unmentioned those which one would rather pass with averted eyes' after 'the highly successful process of development'.

IN SEARCH OF THE IDYLL

The idea of 'day-tripping' and 'weekends in the country' were incredibly alluring for the new motoring middle-classes and the workers who flocked out of cities at the end of the working week. C.E.M. Joad's *Charter for Ramblers* of 1934 described how beer had been replaced by hiking 'as the shortest cut out of Manchester', whose Central Station was an unforgettable mass of 'rucksacks, shorts and hob-nailed boots' every Sunday morning. Joad had led his own railway pilgrimage to Chanctonbury Ring two years earlier, when 16,000 fellow ramblers

'The English countryside was ... the very embodiment of decency, Englishness, national character and national identity.'

enjoyed the communal experience of watching the sun rise over the Iron Age fort. By this time, 100,000 men and women were reckoned to be exploring the countryside on a regular basis through the new leisure crazes of rambling and rock climbing.

The rise of mass car-ownership was the decisive factor in turning the countryside into the nation's playground. From only 78,000 in 1918, there were a million cars on England's roads by 1930 and two million vehicles in total by 1933. The rural idyll was the obvious destination for motorists who were now free to explore England's less-beaten tracks. Motoring had always been seen as a leisure activity, but the main source of pleasure gradually changed from a Toad-like pursuit of speed to a more responsible search for beauty. The rise of motoring accompanied the rebirth of the travelogue; following the earlier tradition of Fiennes, Defoe and Cobbett on four wheels were the likes of H.V. Morton, with his seminal *In Search of England* of 1927, a paean to the countryside he never expected to see again. While delirious with spinal meningitis in Palestine, Morton described the hallucinations of England as 'the only religious moment I experienced in Jerusalem'; what rose up in his mind were not the landmarks of his home city of London, but the wood smoke and elms of the countryside, complete with 'the slow jingle of a team coming home from fields'.

Morton vowed that if he recovered, he would go in search of the England which had been imprinted in his imagination by centuries of conditioning: the England of 'lanes and little thatched villages'. Thankfully, Morton did recover and lived until 1979, but he is still remembered for his 'road trip' in a bull-nose Morris. Morton acknowledged that cheap motoring meant 'more people than in any previous generation seeing the real country for the first time', but hoped those in search of the rural idyll would discover 'a living thing' rather than just a 'pretty thing'.

By the time J.B. Priestley drove out of London on the first leg of his *English Journey* of 1933, the pace of recent change made Morton's England feel like a distant relic. Sir Philip Gibbs marked the Jubilee of George V in 1935 with a state-of-the-nation survey, *England Speaks*, which declared with some relief that despite the 'superficial' damage of industrialisation, the 'soul of England' – the countryside – remained mostly intact. Along with the resurgence of the travelogue, the rise

Lower Shiplake in Oxfordshire, the setting of Lower Binfield – George Orwell's metaphor for 'the green and pleasant land'.

of motoring created a huge market for a new wave of multi-volume guidebooks celebrating the English countryside, such as the county-themed volumes of Arthur Mee's *King's England* series. From 1936, modern artists began editing John Betjeman's more irreverent Shell Guides, which, following Betjeman's ruthless portrait of Slough in 1937's *Buckinghamshire*, increasingly seemed to document the demise of rural England. Betjeman's famous poem of the same year took aim at the town's urban sprawl, which meant there was no longer 'grass to graze a cow', but showing people the harsh reality only seemed to fuel their desire to retreat to the idyll.

The popularity of real rural reportage, such as Adrian Bell's farming trilogy – *Corduroy* (1930), *Silver Ley* (1931) and *The Cherry Tree* (1932) – betrayed an urban England falling in love with the idea of the countryside. Bell had made his own pilgrimage back to the 'soul of England'. A highly educated city-boy – and first compiler of that epitome of urbanity, *The Times* cryptic crossword – Bell had decided to reject polite society and to get his hands dirty as a farmer in deepest Suffolk. By the end of the decade, George Orwell's *Coming Up For Air* portrayed an English everyman, George Bowling, looking to make a similar

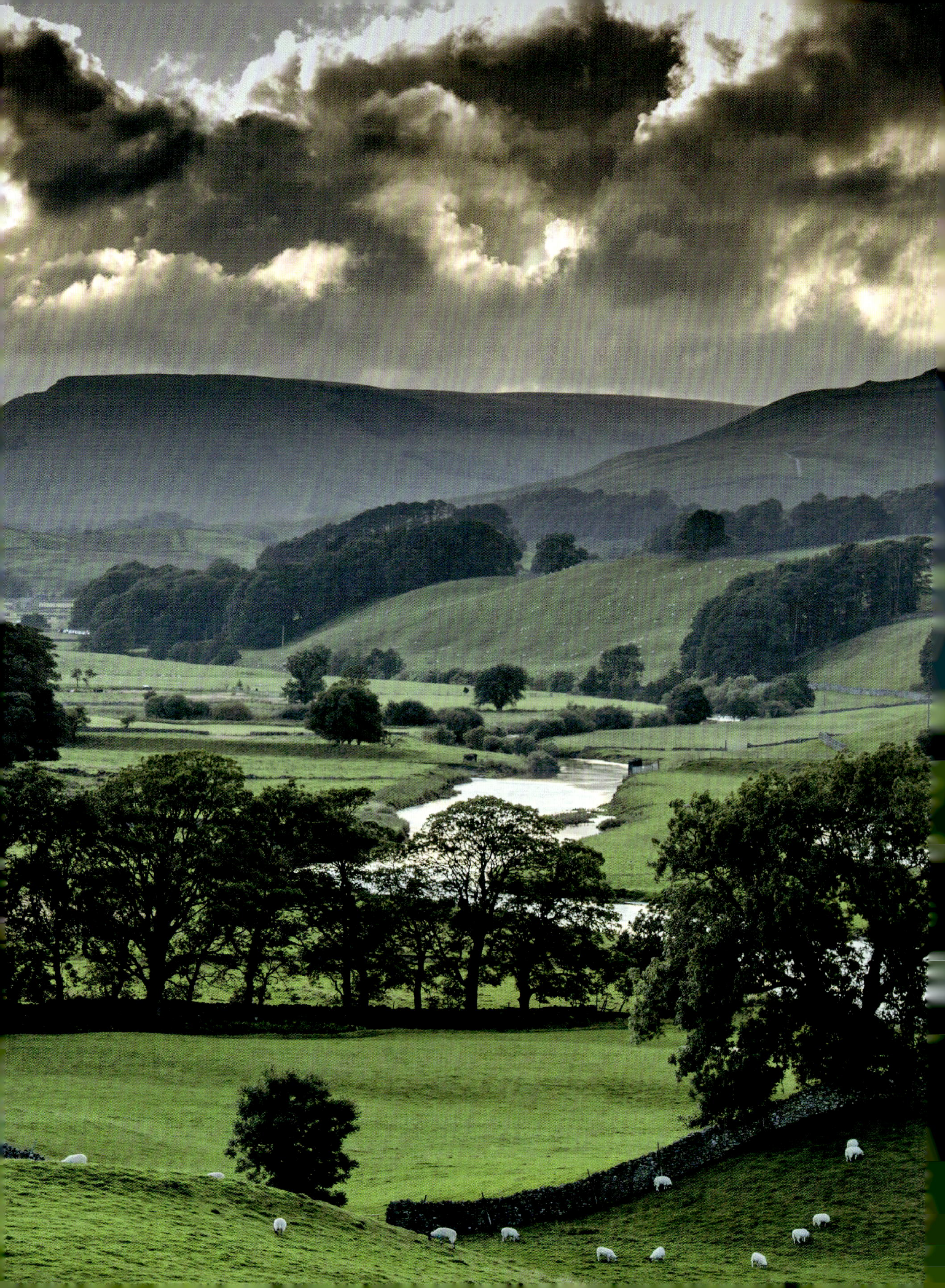

OPPOSITE
Wensleydale in the Yorkshire Dales provides an escape from modern, city living, just as Orwell's 'Golden Country' did in *1984*.

ABOVE
Combe Gibbet in Berkshire remains a popular destination for walkers, motorists and cyclists in search of the rural idyll.

escape to the country. Bowling, overwhelmed by suburban life, found the smell of horse dung triggered his first daydream, his rural childhood, after which he knowingly addressed the reader: 'Is it gone for ever? I'm not certain. But I tell you it was a good world to live in. I belong to it. So do you.'

Upon driving towards his old village of Lower Binfield – Orwell's metaphor for the rural idyll – Bowling had the horrible realisation that all the fields had been covered in houses, seeing only 'ghosts, chiefly the ghosts of hedges and trees and cows'. Was the English notion of the countryside becoming, as Bowling suggests, 'a thin bubble of the thing that used to be, with the thing that actually existed shining through it'? In one sentence, Orwell deconstructs the rural idyll: 'All those years Lower Binfield had been tucked away somewhere or other in my mind, a sort of quiet corner that I could step back into when I felt like it, and finally I'd stepped back into it and found that it didn't exist.'

Despite dashing George Bowling's dream of the rural idyll, there was a happy ending of sorts from George Orwell in *1984*. In his last novel, published in 1949, we find the 'Golden Country' of Winston Smith's recurring dreams actually exists in the dystopian future of Bowling's worst nightmares. Orwell portrays the countryside as a refuge from the telescreens and concealed microphones of the city. Perhaps inspired by 1949's National Parks Act and the rural England's role as a refuge during the war, Orwell gives hope that the countryside will always endure in even the bleakest times, even though the pastures may be 'close-bitten' and the hedges 'ragged'.

That our notion of the green and pleasant land was created during two of the most iconic half-centuries in England's history adds to its enduring appeal; the Elizabethan age planted the concept in fertile ground, while the late-Victorians gave it a moral resonance. If the idea had not been so squarely at the centre of our national character, it is impossible to imagine that we would have succeeded in protecting so much of it, given the pressure to develop. As the pace of modern urban life takes us ever further from the natural rhythms of the countryside, there is no likelihood that we will give up the national obsession of dreaming about escaping back to the country.

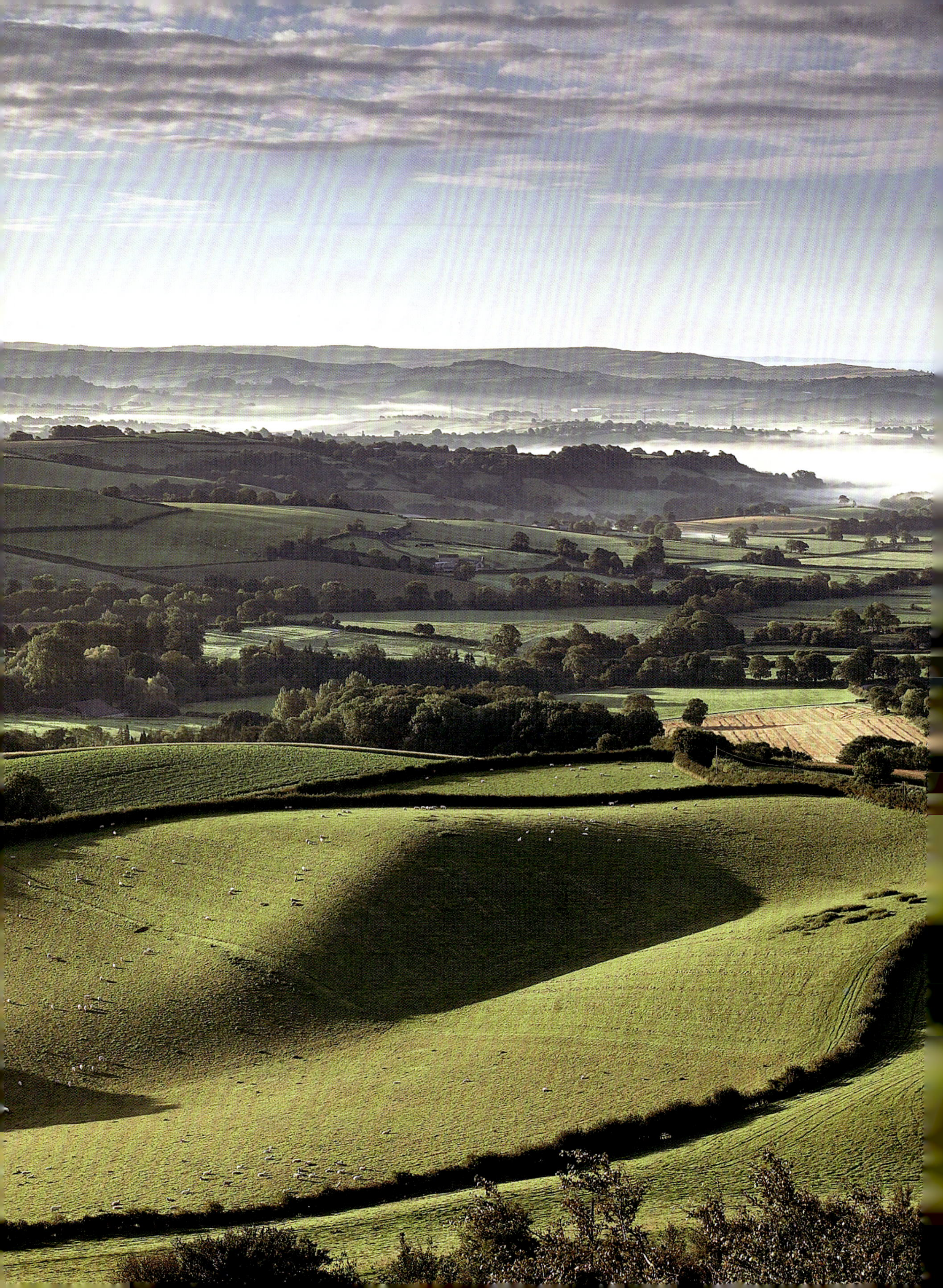

Idea 13

A Countryside Worth Fighting For

introduced by Tony Robinson

We'd never seen a proper bluebell wood before, so on the Sunday after the 70th anniversary of VE Day, we took a trip to the legendary Franks Wood in Surrey. It was stunning – half a square mile of outrageously vivid colour under a canopy of gnarled and ancient trees.

On the way back to the car park my wife said 'That's what we fought the war for, wasn't it?' and although it sounded a bit odd at the time, on reflection I think she hit the nail on the head.

Nowadays we take it for granted that we're stewards of the English countryside and have a moral duty to protect it, but that perception didn't appear by magic – it was nurtured over two World Wars by our artists and film makers, Government propaganda, and in no small part by the CPRE.

IF the rural idyll was firmly established in the English psyche by 1900, the twentieth century tested our resolve to defend it. The countryside was deeply entwined with the ideals of England being defended in both World Wars, and we still think of the 'green and pleasant land' as something that our forebears sacrificed their lives to pass down. Because our countryside was fought for, it feels like our greatest inheritance, and explains why we feel so passionately about protecting it from self-inflicted wounds.

In August 1914, on the brink of the First World War, Churchill wrote – from 'a hilltop in the smiling country which stretches round Mells' in Somerset – of his horror at the prospect of the young men of England being torn away from 'the idle hill of summer, sleepy with the sound of dreams'. In quoting A.E. Housman's *A Shropshire Lad*, Churchill alludes to the powerful hold that the rural idyll had over so many of those who were expected to go and fight for it. Early recruiting advertisements like 'Your Country's Call' made the most of this, asking 'Is this worth fighting for?' of a patchwork English landscape.

WAR POETS AND THE POWER OF PLACE

The poet Edward Thomas was one of many soldiers for whom the war deepened feelings towards the countryside. Despite having no quarrel with Germans and a hatred of nationalism, Thomas eventually enlisted in 1915. When asked exactly what he was fighting for, he picked up a handful of English soil and replied simply, 'For this.' Thomas had struggled to reconcile his love for the countryside with his initial pacifism: 'It seems foolish to have loved England up to now without knowing it could perhaps be ravaged and I could and perhaps would do nothing to prevent it.' These agonising thoughts must have been shared by thousands of young men who went to war, not for King and country, but for the countryside and their right to enjoy it with a clear conscience. 'Something,' Thomas felt, 'had to be done before I could look again composedly at English landscape.' Tragically, he would never get the chance, making the ultimate sacrifice in 1917.

The strength of Rupert Brooke's Gallipoli-bound yearning for the English countryside had brought about in him 'the triumphant helplessness of a lover'. Ivor Gurney wrote home from the trenches to describe his longing for the 'tiny, dear places' he had left behind in rural Gloucestershire – the 'orchards and roads winding through blossomy knolls'. Ford Madox Ford's wartime memoir recalled a vivid dream he had had in the trenches, in which the English countryside appeared to him as a comforting 'nook . . . a valley; closed up by trees . . . deep among banks; with a little stream, just a trickle. . . . You understand the idea – a sanctuary.'

The war poets seemed to be able to articulate the importance of the countryside as a motivational force or source of comfort for the average soldier, for whom Arthur Quiller-Couch's *Oxford Book of English Verse* was staple reading material in the trenches. Its pastoral poetry was intended to provide an antidote to the nationalism of 'Rule Britannia' and allow the

PAGES 130–1
The Marshwood Vale from Pilsdon Pen in Dorset: the kind of rural view that was a huge motivation for those engaged in the defence of the country in two World Wars.

IDEA 13 • A COUNTRYSIDE WORTH FIGHTING FOR

LEFT
The use of a Scottish infantryman in this 1915 poster merely highlights the strong hold such unmistakably English landscapes had over Britain. For many Tommys – and particularly soldier-poets like Edward Thomas – the defence of the countryside was the main motivation for enlisting.

RIGHT
For the soldiers already on the front line a gentler form of propaganda was needed; this 1917 George Clausen poster was designed to keep alive the dreams of returning home to the village.

Tommy to escape to 'a green nook of his youth; where the folk are slow but there is seed time and harvest'. Before the end of the war, the YMCA's poetry anthology, *The Old Country*, was also produced in the hope that the same Tommy, 'in his imagination, can see his village home'. London Underground's marketing supremo Frank Pick commissioned a series of morale-boosting posters depicting idealised rural scenes by artists like John Walter West, and then arranged for copies to be sent to the trenches in the hope that they would 'awaken thoughts of pleasant homely things'.

IN DEFENCE OF THE COUNTRYSIDE

The horrors of war served to heighten appreciation of both the importance and vulnerability of the countryside, leading to growing concern about the creeping effects of industrialisation. In attempting to make sense of the carnage, the inevitable question was raised in people's minds: what had we been fighting for? The countryside seemed as good an answer as any, and offered something to cling on to for a shattered nation. So, in the run-up to the Second World War, the realisation that invasion was a very real prospect made that countryside seem more vulnerable than ever. The earliest preparations highlighted that perhaps the greatest threat was from 'friendly fire', with the *Manchester Guardian* of 1937 noting 'the fatal fascination

WHY BOTHER ABOUT THE
GERMANS INVADING
THE COUNTRY?

INVADE IT YOURSELF
BY UNDERGROUND AND MOTOR-'BUS

EASTER · 1915

IDEA 13 • A COUNTRYSIDE WORTH FIGHTING FOR

OPPOSITE
Only the marketing gurus of the Underground could see the war as an opportunity to promote the countryside; a carefree approach to home front propaganda from 1915.

ABOVE
Frank Newbould's classic 1942 poster utilised the timeless motif of the hill shepherd as the last remaining link to the traditional idea of the pastoral life. The landscape was later found to be a composite of two separate views, one of the Birling Gap near Eastbourne and one from the hills above Lewes.

which draws our Defence departments, when they are in search of sites for new depots, nearly always to scenes of great beauty or historic interest'.

By 1939, Prime Minister Neville Chamberlain had appointed CPRE as consultants in the process of locating aerodromes and arms factories to help ensure that 'other national interests, especially our finest unspoilt scenery, shall receive due consideration before sites are selected'. The sentiments were even echoed by Mrs Chamberlain in May 1939, in urging that 'in spite of all the difficulties of the moment the splendid work of the CPRE should go on'.

A leading article in *The Times* of the same month welcomed CPRE's advisory role, and warned against 'letting beauty go hang because the nation must have more guns and aeroplanes':

> The end of all this preparation is to preserve in permanence all that makes life worth living in our country; and it would be futile to let the hurry and heedlessness of the hour destroy any portion of what we are aiming to preserve for all time.

As a First World War veteran, and someone who returned to domestic military duties in 1940, CPRE's Captain Herbert Griffin was well qualified to argue that 'young men and women will join up and perform their National Service in better heart if they are satisfied that powerful forces are at work to prevent such places as the Peak District and the Lake District, of which they are passionately fond, being filched from them'. Griffin would certainly have felt vindicated when in 1939, one young soldier wrote to CPRE's office at 4 Hobart Place in Victoria, begging them to continue their campaign for National Parks because 'whatever damage might be done by enemy action, I do not wish to return and find destruction of our own creation'.

RECORDING VANISHING ENGLAND

The question of whether our own carelessness in development was a greater threat than enemy action was raised by a Domesday scheme initiated by the National Gallery director Kenneth Clark in 1940. The plan was for underemployed artists to produce watercolours of 'places and buildings of characteristic national interest exposed to the danger of destruction by the operations of war'. Upon hearing of the scheme, Frank Pick suggested that Charles Knight should make paintings of the South Downs, which seemed at risk of imminent invasion – either from Nazis or garden city developers. Pick had been working with CPRE throughout 1939 in the campaign to protect 700 acres of the South Downs around Ditchling from a 'garden city' of 10,000 inhabitants.

After commissioning Knight, Pick enlisted CPRE's help in suggesting other locations for a 'Record of Disappearing Britain'. When sixteen of CPRE's county branches suggested landscapes as likely to be destroyed by irresponsible builders as Nazi bombs, Clark was forced to direct his painters to prioritise 'fine tracts of landscape which are likely to be spoiled by building developments'. The art critic Herbert Read summed up the importance of the exercise in 1941:

> Better than the wordy rhetoric of journalists or politicians, it shows us exactly what we are fighting for – a green and pleasant land, a landscape whose features have been moulded in liberty, whose every winding lane and irregular building is an expression of our national character. There will be little point in saving England from the Nazis if we then deliver it to the jerry-builders and the 'development corporations'.

The artist John Piper later looked back on his work for *Recording Britain* and felt that 'the fear of losing the precious buildings and sites of England' gave them 'great visual and emotional importance'. For Piper, the war had helped artists re-learn how to appreciate the countryside: 'to look and see the point of things that they had totally discarded', like the 'beauty spots' they had been 'taught to shun by Roger Fry and Clive Bell'.

This 1940 watercolour of a Derwentdale farm was captured for the Recording Britain *project before the proposed Ladybower Reservoir (featured on this book's cover) flooded the valley. Although the CPRE Sheffield and Peak District branch commissioned the painting to preserve a record of the 'exceptionally beautiful area', they supported the reservoir in the belief it could be built to harmonise with the character of the Dale.*

THE GOLDEN ACRES OF HOME

Wartime propaganda films like 1940's *Our Island Fortress: There's A Land, A Dear Land* used rural imagery to represent England's 'freedom and glorious heritage'. In the autumn of 1942, aerial footage of 'the good golden acres of home' was used by *Our Record Harvest* in a reminder that 'our land has given us richly of its bounties' and 'behoves us all to use it thankfully and not abuse it thoughtlessly'. In the same year, the recruitment drive was given a patriotic push by the War Office's commissioning of a series of images from the travel poster artist, Frank Newbould. Using all his experience of selling rural England to the urban masses, Newbould obliged with four posters showing 'Your Britain' and demanding 'Fight For It Now'. No matter that the posters – two village greens, Salisbury Cathedral and the South Downs – were unashamedly rural, southern, and unlikely to represent the Britain of more than 10 per cent of those joining up, Newbould's posters were pure nostalgia for a disappearing world. By far the most famous poster showed a lone shepherd and his dog high up on the South Downs. When chosen by former CPRE Chairman Peter Waine for *Country Life*'s 'My Favourite Painting' column, their art critic John McEwen suggested that

Evelyn Dunbar's *A Land Girl and the Bail Bull* from 1945. The Second World War brought thousands of people into close contact with the countryside for the first time.

the image 'reflected Churchill's view of the war: Britain as the immemorial defender of tranquil liberty opposed to German industrialism run amok'.

In 1943, with the outcome of the war still uncertain, the naturalist Peter Scott – son of Scott of the Antarctic – summed up the importance of the countryside to our comprehension of what we were fighting for. During his time in the Royal Navy, he became aware that England 'for most of us is . . . the English countryside'. As his destroyer patrolled around Devon's cliffs at dawn one day in 1940, Scott's mind was filled with images of 'wild moors and ragged tors . . . narrow winding valleys with their steep green sides – the countryside we were so passionately determined to protect from the invader'.

A few months earlier, John Betjeman, exiled from England as the UK's press attaché in Dublin, confessed that his enforced absence had made his heart grow fonder – not just of the architecture that he spent most of the 1940s eulogising, but of the countryside of his pre-war home. Betjeman's broadcast on *England Revisited* revealed that he had become teary-eyed at 'the sight of moonlight on a willow stump covered with ivy' simply because it reminded him of a willowy brook back in Berkshire. His only consolation was the thought that the stars shining over Dublin 'were also shining on my home village'. The depth of feeling for the countryside provoked by the war was extraordinary considering, as in Betjeman's words, his generation had been taught to 'avoid the sloppy and sentimental'.

THE NATION CLAIMS THE COUNTRYSIDE

By the closing stages of the war, a greater proportion of the population had been exposed to rural England than at any time in the preceeding century. Nearly 100,000 young women joined Lady Denman's Women's Land Army, while around 2 million children and mothers left cities to avoid the expected bombing in September 1939. A thousand more of our military personnel were trained or billeted in the countryside, with 'The Few' of the RAF observing perhaps the most humbling view of its fragility yet seen by human eyes.

In addition to these human experiences, the weight of propaganda and poetry that romanticised the country during the war had a huge impact. Not even popular music was immune. The first two months of the Second World War saw 200,000 copies sold of a Vera Lynn tune promising 'There'll Always be an England' – 'wherever there's a cottage small beside a field of grain'. The most popular song at the end of the war, 'We'll Gather Lilacs', featured Ivor Novello crooning about the joys of walking along English country lanes in springtime. This growing familiarity with rural England was perfectly captured through the character of a London soldier in the 1944 film *A Canterbury Tale* exclaiming that he 'never realised there was a countryside before the war'.

The Second World War helped the public feel a sense of ownership over the countryside, and the National Parks campaigner Sir Norman Birkett felt that new legislation to protect it would be a form of compensation for 'so much sacrifice and so much suffering so nobly borne'. Hugh Dalton, the Chancellor who paid for this post-war settlement for the countryside, acknowledged the importance of safeguarding 'opportunities for health and happiness, companionship and recreation, in beautiful places . . . which Hitler never scaled. Let this land of ours be dedicated to the memory of our dead, and to the use and enjoyment of the living for ever.'

Our modern, country-loving form of Englishness was forged in the white heat of war. Even the unofficial national anthem of England was created when Blake's description of the 'green and pleasant land' was set to music by Hubert Parry in 1916. In times of national crisis, it was the countryside that was the great motivating and unifying force. People were willing to fight for the hope of a better country symbolised by a rural past, which compared to the deprivations of modern urban living did seem idyllic.

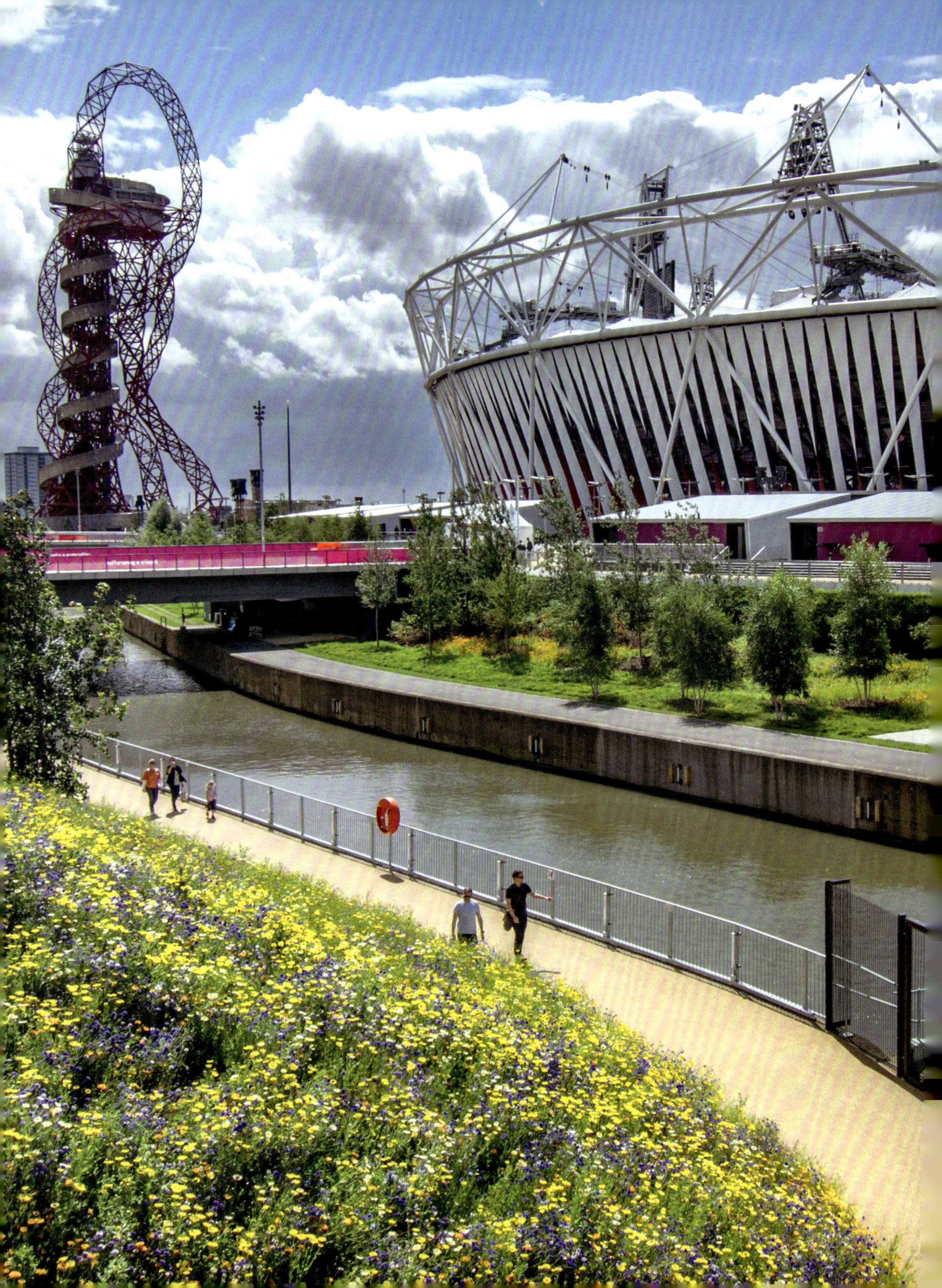

Idea 14

Urban Regeneration

introduced by Terry Farrell

From the nineteenth century we had inherited unplanned growth, particularly in chaotic opportunistic industry and infrastructure layouts, and vast turmoil in ever-expanding mass housing. A brief period of people moving away from the problems of cities to new towns, garden cities and suburban sprawl was followed by a determined and eventually highly successful retrenchment in the city centres.

Capitalising on patterns of vacancy and opportunity, vast areas of ports, goods yards, former power stations and gas works have now been redeveloped. In particular, from Colin Buchanan's time onwards, the motorcar and urban life have become better balanced, with pedestrianisation, traffic management and congestion charging.

The best way of protecting the countryside is to make urban living as attractive an offer as possible, and today urban life is resurgent. People have returned to live, work and play in our post-industrial urban areas. It is as much an urban revolution as a regeneration.

Every home built on a previously developed 'brownfield' site saves a piece of a green field. Around 150 square miles of this derelict land have been developed for housing since CPRE helped make 'brownfield first' Government policy in 1995, saving enough green fields to accommodate seven Southamptons. The idea of rebuilding in inner cities to improve living conditions has been around for centuries, but the notion that improving urban England could help save the countryside seemed completely counter-intuitive. So it is fitting that it may have been inspired by one of England's greatest lateral thinkers – the economist John Maynard Keynes.

When Clough Williams-Ellis was assembling his 1937 collection of essays, *Britain and the Beast*, he placed Keynes' radical article on 'Art and the State' first in a formidable running order that included E.M. Forster, C.E.M. Joad and Patrick Abercrombie. The book was designed to promote and encourage the work of CPRE and the National Trust, with ideas from the finest minds of the interwar generation. Keynes understood that preserving countryside from development in order to realise its value for 'health, recreation, amenity, or natural beauty' reduced the amount of land available for new housing and other development. His solution was to look at urban areas where 'space is at present so ill used that a larger population could be housed in modern comfort on half the area or less'.

Uncannily anticipating the future regeneration of London's South Bank, Keynes wanted to demolish 'the majority of the existing buildings on the south bank of the river from the County Hall to Greenwich, and lay out these districts as the most magnificent, the most commodious and healthy working-class quarter in the world'. Keynes suggested that if inner cities across England could be transformed in this way it would 'enrich the country and translate into actual form, our potentialities of social wealth'. He argued that civic pride in our cities was vital for social unity, calling for the 'comprehensive rebuilding at the public cost of the unplanned, insalutary and disfiguring quarters of our principal cities'.

The Second World War, and particularly the Blitz of London, opened other eyes to the potential Keynes had already seen; somewhat bizarrely, it was a 1941 advert for Pears soap called 'The Town of The Future' that first imagined how the post-war reconstruction could correct the mistakes of the past, where 'towns and cities have straggled and sprawled, capturing parts of the countryside with the same inevitable disappointment as the caging of a wild bird. The town of the future will be erect and compact.' The poster makes more sense when one realises that in 1920, Pears was acquired by the Lever Brothers, who had recently completed their model community. They may well have felt that Port Sunlight offered a template for the rebuilding process, providing workers' housing within a mix of contrasting architectural styles, green spaces and public buildings.

PAGES 140–1
The Olympic Park in Stratford, London, featuring the *Orbit* sculpture created by Sir Anish Kapoor, a supporter of CPRE's campaign for urban regeneration.

IDEA 14 • URBAN REGENERATION

Colin Buchanan's influential *Traffic in Towns* report of 1963 identified the need to protect the countryside by rejuvenating towns and cities as pleasant places to live, as illustrated here by Kenneth Browne's vision of a car-free Newbury.

A MISSED OPPORTUNITY

Also seeking a silver-lining to the damage of war were the Labour MP Gilbert McAlister and his wife Elizabeth, who set out their plan to rebuild England's *Homes, Towns and Countryside* in 1946: 'Never before in the history of mankind has been presented such a magnificent opportunity to provide pleasant homes for all in beautiful towns and cities, set against a thriving and unspoiled countryside.' Unfortunately, immediate rebuilding on bombed sites was very costly and could not be done anywhere near fast enough, so new towns were increasingly seen as the logical way to re-house the urban population. CPRE opposed the New Towns Bill of 1946, making it clear that the 'building of new towns should not be undertaken' on virgin countryside until the full potential of suitable 'war sites' had been realised.

In Parliament, the previous Planning Minister William Morrison pledged to 'oppose this Bill if I thought that the creation of these new towns was to be at the expense of re-beautifying and reconstructing the old towns'. Recalling the tea party in *Alice's Adventures in Wonderland* where 'the Dormouse upset the teapot and they all moved round to a clean place on the tablecloth', Morrison made an eloquent plea for urban regeneration decades before the term was invented:

> I would not like our local authorities to be deluded by this Bill into thinking that they can let decay crumble away the interiors of their own towns and then walk out on to more and more clear land to rehouse the population, which ought to have been housed in decency and dignity inside the old town.

Viscount Hinchingbrooke agreed, fearing that 'we may be only too easily exchanging that policy of unplanned ribbon development for a policy of planned cannibalism'. Instead of invading the countryside with new towns and 'leaving behind great masses of derelict property', he was strongly in favour of 'remodelling existing towns'.

Meanwhile, Morrison's successor Lewis Silkin – who would become a hero of the countryside after the 1947 Town and Country Planning Act – found himself vilified in rural Hertfordshire when work began on Stevenage New Town in 1946. Locals created 'SILKINGRAD' signs for the train station and the tyres of his ministerial car were let down while he tried to convince residents of the 'old town' that 'people from all over the world will come to Stevenage' new town. CPRE successfully opposed two new towns in Cheshire on the grounds of loss of productive agricultural land, and their public inquiry evidence in 1949 also saved 770 acres of farmland from becoming part of the new town of Bracknell.

These campaigns made the loss of farmland a hot topic; in March 1949 Dudley Stamp – the Director of Britain's Land Utilisation Survey – castigated the policy of building New Towns on good farming land 'in a world where land exhaustion and increasing population mean that per capita world supplies are dwindling'. After weeks of debate in the press, Stamp warned that there were 'no reservoirs of rich virgin lands awaiting the first turn of the plough as there were 80 or even 50 years ago'. He concluded by quoting Earl Radnor: 'I am not convinced that our descendants will appreciate having been provided with "new towns to starve in".'

NEW VISIONS FOR OLD CITIES

A decade of new town building produced the inevitable reaction. In 1961, the architect Lionel Brett launched a new movement berating new towns for their lack of 'compactness, convenience, neighbourliness, excitement', as well as their inefficient use of land. Brett was tapping into the radical new ideas of the American writer Jane Jacobs, whose seminal *Death and Life of Great American Cities* was published that year. Though it would not become truly influential for another decade, Jacobs' 'new urbanist' thinking held that inner cities should utilise derelict industrial buildings for high-density housing and businesses, creating vibrant, pedestrian-friendly neighbourhoods. At the same time, urban quality of life was being eroded by a rapid increase in traffic that threatened to thwart any hopes of making urban renewal fashionable.

Two years after Jacobs, future CPRE President Colin Buchanan produced his landmark report on *Traffic in Towns* for the Government, realising that society had to confront its addiction to the car if cities were to remain viable places to live and an alternative to building on green fields. He offered two choices: reduce car use and enhance quality of life by improving public transport and pedestrianising city centres, or simply allow cars to dominate by building motorways right through the heart of historic cities. The second option was an extreme scenario, which Buchanan never promoted but which was seized upon by some planners to justify madcap traffic schemes, such as the demolition of Piccadilly Circus for an urban motorway – a proposal that limped on until the 1970s.

In a 1964 Lords debate on urban redevelopment, Lord Conesford praised the Buchanan report for making it clear that 'the preservation of the countryside depends on the possibility of a civilised life in the towns'. Future CPRE Chairman Lord Molson agreed that 'urban redevelopment is a matter of urgent necessity' in order to avoid Buchanan's worst nightmare of 'an urban sprawl extending over almost the whole of the countryside'. Ultimately, the long-term influence of Buchanan helped to make England's cities much friendlier places for people; together with growing appreciation for Jane Jacobs' ideas, this laid the foundations for an urban renaissance of city-centre living.

INNER CITY BLUES

New towns and the relocation of industry had exacerbated the rapid decline of industrial city centres; the UK's major cities shed 35 per cent of their population between 1951 and 1981 in what CPRE called the 'urban exodus'. The feeling that investment in new towns was at the expense of established cities gained support to the point that in 1976 Labour's Environment Secretary Peter Shore argued the case for diverting 'the country's resources towards inner-city areas'. Increased State investment in urban areas followed – at the expense of funding for new towns. By March 1986, Prime Minister Margaret Thatcher was able to assure CPRE President David Puttnam of the Government's commitment to look for alternatives to green fields for new building:

> We are seeking the right balance between the need for development and the interests of the environment. Our priority is to re-use previously developed land, particularly in urban areas, and we have made substantially increased resources available to reclaim derelict land for redevelopment.

The BBC's decision to relocate a large part of its operation to Salford shows the potential for urban regeneration to promote more even growth and create vibrant new communities.

This shift in attitudes helped to vastly increase the proportion of housing on previously used (brownfield) sites from 38 per cent in the early 1980s to 45 per cent by 1987. Environment Secretary Nicholas Ridley found it 'remarkable that we have managed to use old land for 47 per cent of all development', but aspired to do even better in the future. CPRE agreed that more could be done; in his Foreword to their 1987 book *England's Glory*, the novelist John le Carré said 'the CPRE questions, as no partisan body can . . . the point of allowing the Titans of the housing lobby to eat up more and more greenfield sites while the inner cities, with all the wretched social consequences we know about, fall into squalor and decay.' In 1992, David Puttnam's successor as CPRE's President, Jonathan Dimbleby, appealed to television viewers to 'think of those great wastelands inside our towns and cities crying out for redevelopment and renewal'.

By the mid-1990s, CPRE's campaigning had encouraged Ridley's successor John Gummer to set a target for at least 60 per cent of new homes to be built on brownfield land. The Labour Government carried on the good work; in 1998, Environment Secretary John Prescott called for 'an urban renaissance in Britain' to ensure 'that we hand on a green and pleasant land to future generations'. Prescott appointed the leading architect Richard Rogers to lead an Urban Task Force including CPRE's Tony Burton, leading to official 'brownfield first' policies to exhaust derelict sites before greenfield sites could be considered.

RESTATING THE CASE FOR BROWNFIELD

'Brownfield first' has been one of the most successful and environmental, social and economic policies ever seen, anywhere. By 2008, an incredible 80 per cent of new homes were being built on brownfield sites, saving a huge area of countryside while transforming derelict sites like Stratford's Olympic Park. Thriving parts of Manchester, Birmingham and Liverpool have also been transformed by imaginative urban regeneration over the last twenty-five years. Unfathomably, the Coalition Government felt 'brownfield first' was holding back the economy and housebuilding, and scrapped it in 2010. As a result, the proportion of houses built on brownfield fell back to 60 per cent just as 700,000 homes were earmarked for Green Belts and open countryside.

In response, the then CPRE President Sir Andrew Motion launched a Charter to Save our Countryside – backed by Rogers, le Carré, Dimbleby and Terry Farrell among many others – calling for a return to 'brownfield first'. At the same time, John Gummer (now Lord Deben) summed up the choices facing the country: 'Building on green fields is the lazy way to sacrifice our future. Let's concentrate on recycling already-used land.' Within a year, the Chancellor George Osborne was using his Mansion House speech to pledge to 'remove all the obstacles that remain to development on brownfield sites', to preserve the countryside he described as 'part of the inheritance of the next generation'.

But with the Chancellor's ambition for brownfield housing limited to 200,000 homes, CPRE launched new research at the end of 2014 showing there was enough brownfield land for 1 million homes, and that such land is becoming available faster than it is developed. Crucially, around half of these sites are in London, the South East and the East of England, where demand for new homes – and the threat to Green Belt and open countryside – is greatest. In his foreword to the report, Richard Rogers reiterated the link between urban regeneration and rural protection by calling for new settlements within 'our towns and cities, not on inaccessible and unsustainable green field sites'.

The Coalition subsequently announced that their two flagship 'garden cities' would be built on brownfield sites in Ebbsfleet and Bicester. And with the new Conservative Government pledging to honour its manifesto commitment to prioritise brownfield using a £1 billion fund, there are promising signs that we could see a lasting urban renaissance and an end to the misguided, short-term solution of building on the countryside.

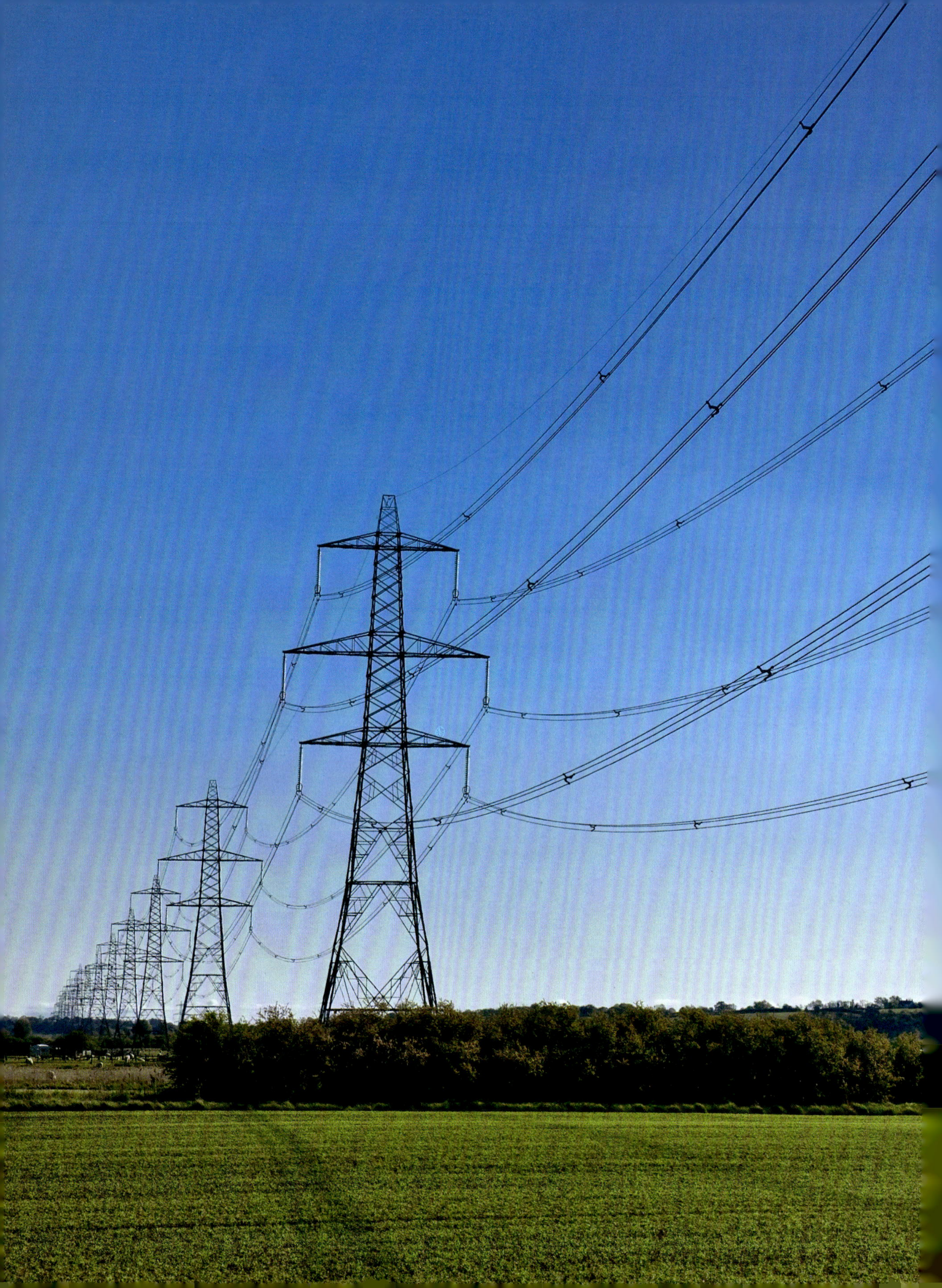

Idea 15

Going Underground

introduced by Chris Baines

Although some people may see a certain bold elegance in the lines of pylons that carry power lines across the landscape, there can be little doubt that our most precious scenery would be much improved without them. In 2013 the Government's energy regulator Ofgem approved a programme to reduce the visual impact of pylons and power lines in the worst affected stretches of National Parks and Areas of Outstanding Natural Beauty.

This Visual Impact Provision policy has brought a range of contrasting partners together, each playing to their strengths. Landscape consultants ranked the respective scale of visual impact in each affected landscape; the National Grid's planners and engineers assessed the technical challenges of undergrounding through sensitive places; local specialists shared their ecological and archaeological knowledge; and local communities brought their own judgement to bear.

With all that excellent support, the panel of individuals representing stakeholders such as CPRE, the National Trust, the Ramblers' and several other countryside charities and Government agencies was able to agree the first landscapes that should lose their pylons – in Snowdonia, the Peak District, the New Forest and the Dorset Downs.

Other regulated industries also have an impact on our landscapes. Water and rail companies are obvious examples. If their regulators follow Ofgem's lead, their funds could pay for sustainable water catchment management and habitat restoration alongside rail corridors. In time, who knows, the health regulators may even insist that pharmaceutical companies help to finance healthy access to fresh air and exercise in natural surroundings.

The fledgling electricity network of the 1920s was a symbol of hope for remote, rural communities around England. But as powerlines began to seriously intrude on landscapes, CPRE began recommending that they should be buried underground, particularly 'through commons, playing fields, National Trust properties and areas of great beauty'. In February 1933 they published research on the costs involved, declaring that although 'diving underground' was costly, the alternative was often 'damage to amenity or beauty which might be intolerable'. Later that year, Stephen Spender's poem, *The Pylons*, summed up the impact of such huge structures on places where 'the secret of these hills was stone, and cottages':

> Now over those small hills they built the concrete
> that trails black wire
> Pylons those pillars
> Bare like nude, giant girls that have no secret

PAGES 148–9
Pylons bestriding the countryside of East Sussex. So much more of our countryside would have been intruded upon by such structures if not for campaigns for 'undergrounding'.

THE WIDENING WIRESCAPE

The Lancashire branch of CPRE had worked to ensure 95 per cent of the county's rural network had been put underground, but by 1936, a 61 per cent increase in the cost of cable meant the pylons began springing up. In 1937, the situation prompted a *Manchester Guardian* leader bemoaning that although 'the Council for the Preservation of Rural England is sending out healthy branches from which developments in the countryside are being watched . . . distributing companies have wide powers, and the Minister is not sympathetic'. In the meantime, the Friends of the Lake District were representing CPRE and the National Trust in a high-profile campaign to bury a line from Keswick to Honiston quarry. A Commons debate on the issue that December saw Jack Lawson MP remind the Chancellor that 'there are hundreds of thousands of skilled men who could do this [undergrounding] work, and who are now being paid for doing nothing'.

In 1949, after the war-interrupted decade, CPRE's Herbert Griffin began to argue that as 'visitors from towns come each year to see the beauty of the scenery – which they expect to find unspoiled – it is reasonable that the towns should contribute' to a National Fund for undergrounding. CPRE secured some major undergrounding victories in the 1950s, saving the highest points of the Malvern Hills, Holy Island and the picturesque Lancashire hamlet of Abbeystead from pylons. But the announcement of plans for a new 'supergrid' of even bigger pylons even caused the Chair of the Central Electricity Generating Board, Sir Christopher Hinton, to suggest that only CPRE's plan for more undergrounding could halt the alarming trend towards 'wirescapes'. Leading photographer Eric de Maré complained that 'civilisation should come before cost accountancy . . . so let us lay those beastly wires under

IDEA 15 • GOING UNDERGROUND

Fascinating item from a National newspaper, last week: "With intelligent use of space and modern methods of blending highways into the scenery, the beauties of the countryside will always be preserved." Wanna bet?

Giles sent this signed cartoon to CPRE Gloucestershire in the mid-1960s as a mark of appreciation for their anti-pylon campaigning.

the ground and out of sight. It will be a cultural gain of immense value.' A *Times* editorial on 'The Widening Wirescape' conceded that overhead powerlines were 'out of keeping, in scale and material, with their surroundings', and endangered 'the sense of loneliness and remoteness for which modern civilised man often yearns'.

AGE WITHOUT A SOUL?

Building on the sympathetic climate, the Friends of the Lake District and CPRE Northamptonshire convinced their local electricity boards to bury overhead lines in some of their most attractive landscapes. By January 1964, plans for a 15-mile line of the largest new pylons through the recently designated South Downs Area of Outstanding Natural Beauty (AONB) seemed to signal an end to the steady progress. A new ally emerged in John Betjeman, who highlighted the folly of comparing pylons to windmills in a letter to *The Times* that August:

> Windmills could never be put underground. Windmills never marched in straight lines from a central generating station. Windmills were handmade and not all of a pattern. It is sentimental to glorify pylons. We all really know why pylons are allowed to industrialise and change the character of downs and modest agricultural landscape. The reason is money. Since we live at a time when money is regarded as more important than aesthetics we must expect the electricity undertakings

to continue to ruin country districts with overhead wires and cables.

Betjeman's poetic riposte came in the following year's 'Inexpensive Progress':

> Encase your legs in nylons,
> Bestride your hills with pylons
> O age without a soul

A major success of 1968 – when CPRE Gloucestershire convinced the Minister to stop a line of pylons from crossing the Severn at Sharpness 'and adding yet another electrical eyesore to the hard-pressed Cotswolds' – proved to be a rare victory, and the overhead network grew steadily over the next thirty years. By 2003, a survey by the industry regulator Ofgem found that 75 per cent of people wanted to see more cables laid underground, with 89 per cent of people particularly concerned about pylons in National Parks and AONBs. The survey also showed that 40 per cent of consumers would be prepared to pay more (£30 a year on average) for their electricity to see pylon lines undergrounded.

A BREAKTHROUGH FOR PROTECTED LANDSCAPES

In 2006, after campaigning from the Friends of the Lake District, the energy regulator Ofgem included an allowance for undergrounding overhead lines in designated landscapes, meaning cables could be undergrounded in iconic locations like Nelson's village of Burnham Thorpe in the Norfolk Coast AONB. By 2009, almost 7 miles of powerlines had been undergrounded in the Lake District, while the Dedham Vale AONB buried a 1¾-mile line between Constable's birthplace of East Bergholt and Dedham, restoring the scenes of the painter's childhood for a cost of just £1.9 million (easily offset by an annual income of £40 million from tourists seeking out these timeless views).

In 2007, the incoming CPRE President Bill Bryson addressed the economics of a pylon problem, which too often left him feeling as if he had 'wandered onto a set from *War of the Worlds*':

> In 1986, at the time the electricity companies were being privatised, *The Economist* magazine calculated that if all the electricity generating companies were required to devote one half of one per cent of their turnover to burying overhead cables, we would be able to bury 1,000 miles of them every year. There are 8,000 miles of high-voltage powerlines in this country, so they would all be buried now.

Schemes to bury powerlines in 'Constable Country' have helped restore the Stour Valley's landscape to a level of uncluttered beauty the painter would have been familiar with.

Continued lobbying from CPRE convinced Ofgem to create a £500 million fund (costing bill payers just 22p per household) to bury 15 miles of lines in a shortlist of iconic landscapes announced in 2014. But while welcome progress has been made on existing lines – including a remarkable 41 miles of overhead lines buried in the south-east and east of England since 2005 – more action is needed to make undergrounding the default option for new lines, and especially those needed to feed the ever-increasing supply of renewable energy into the grid. The environmentalist George Monbiot has argued that unless these connecting lines are buried, their unpopularity could stall the whole renewables programme:

> Where new powerlines are built they must go underground. If they can't go underground, they shouldn't be built. If we are not against pylons marching over stunning countryside, what are we for?

Elsewhere in Europe, Denmark takes pride in minimising the environmental impact of their energy network; despite needing to connect the world's largest percentage of wind power, they are planning to underground 75 per cent of their electricity grid. Germany has managed to put 80 per cent of its 1.1 million miles of electricity lines underground without sacrificing economic growth or security of supply. In preparation for the 2012 London Olympics, Lord Coe argued that 'the undergrounding of powerlines will provide an uncluttered landscape against which the Games can be staged'. The resulting removal of fifty-two pylons from the Stratford site should inspire us to take the same enlightened attitude in rural areas. Undergrounding has saved some of England's most celebrated views and, with more support, could play a huge role in de-industrialising our countryside and preventing our landscapes from becoming wirescapes.

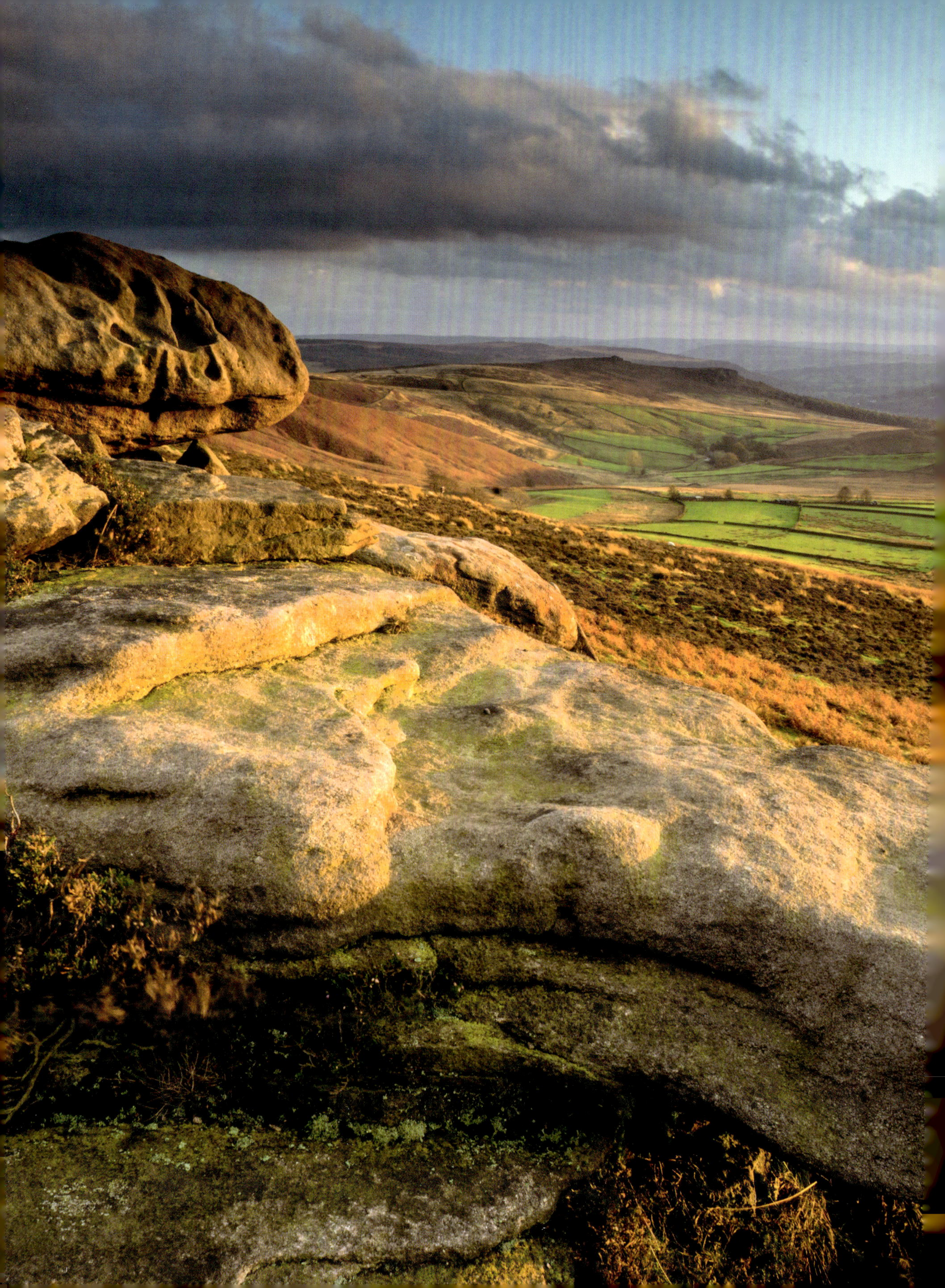

Idea 16

Anti-litter

introduced by Joan Bakewell

All my life I have been moved by the British landscape, from the bleak grandeur of Cumbria and the Scottish Highlands to the curving softness of the Downs, the reedbeds and heathland of its estuaries and the golden sands of my childhood holidays. . . . All this, every bit of it, is now increasingly threatened by the same enemy: litter.

The pioneers of the anti-litter campaign realised that everyone suffers when someone thoughtlessly tosses rubbish from a car window or dumps the remnants of a picnic in a sandy cove. One wonders how much greater the problem would be today without the decades of heroic work from volunteer litter pickers and campaigners celebrated in this chapter. The earliest anti-litter campaigns were right to focus on education. I think we have to teach children to be tidy so that they grow up knowing it's wrong to litter.

Just as it was in the 1920s, the message is: take it home. Then those rolling uplands, the swathes of riverbanks, the curving beaches will be as good as they look in the brochures and magazines, which made us want to see them in the first place. Litter is a blight we can defeat. And we must.

LITTER had always been regarded as an urban phenomenon until, in 1923, Eleanor Rawnsley gave a speech on 'the pollution of the Lake District by the rubbish and litter deposited there by trippers who show indifference to the beauties of our national heritage'. *The Times* suggested that the 'free scattering of miscellaneous litter' was becoming 'the curse not only of the choicest places of natural beauty, but of many humbler scenes as well'. Within two years, King George V himself was publicly complaining of 'the unsightly litter which in so many cases disfigures our parks and other places to which the public resort for recreation and amusement'.

The open spaces campaigner Sir Lawrence Chubb backed the King and suggested that 'local watchers or Boy Scouts and Girl Guides perambulate the popular resorts on Saturdays and Bank Holidays, reminding picnic parties to gather up all their refuse'. He also noted the contribution of modern packaging to the growing problem, saying it 'would be useful if the purveyors of ice-cream and the manufacturers of cigarettes and tobacco could be persuaded to print on the cartons and containers a request not to throw them away in public places'. Taking up this theme, *The Times*, perhaps somewhat guiltily, noted that 'newspapers are printed and read in ever increasing numbers', and that 'many more cigarettes and chocolates in packets are sold than formerly'. Patrick Abercrombie's *The Preservation of Rural England* – emerging the year after the King's intervention – noted that children were actively encouraged to litter, and that one 'recent advertisement of toffee actually showed a small rotund Scots boy with a tin in his hand declaring that he would not return "'till the Toffee's in my tummy, and the tin is in the burn"'.

THE RISE OF THE CAMPAIGN AGAINST LITTER

As well as highlighting the threat to the countryside from 'the litter left, without excuse, by townfolk', Abercrombie won support from the Women's Institutes for his suggestion that rural councils must enforce anti-litter bylaws and provide waste collections to cut down on dumping in hedgerows and ponds.

In August 1929, CPRE's Sheffield and Peak District branch organised a schools litter pick at Stanage Pole outside Sheffield. A few weeks later, the accumulated litter was placed in a glass case as a display at Cutlers' Hall, Sheffield, entitled *What Sheffield Left at Stanage Last Bank Holiday* as part of a CPRE exhibition – possibly the first instance of 'conceptual art' in Britain, beating Hirst and Emin by sixty years! The *Sheffield Daily Telegraph* was not impressed, describing it as 'an indescribable heap of filthy paper, empty fruit tins, broken bottles, cigarette and chocolate wrappers, matchboxes, cigarette ends and other litter'. Nevertheless, over 5,000 local school children were taken to the exhibition and addressed by CPRE members on responsible behaviour in the countryside.

By 1930, the anti-litter campaign was in full swing, with the West Sussex Federation of the WI even selling over 9,000 copies of the sheet

PAGES 154–5
A pristine looking Stanage Edge in the Peak District: scene of a historic litter pick in 1929, and a symbol of the litter-free countryside we'd all like to see and enjoy.

IDEA 16 • ANTI-LITTER

Girl Guides get stuck in at an Anti-Litter League clean-up in the early 1930s.

music to Mr Manning's 'Anti-Litter' song. CPRE and the WIs began joining forces in setting up local 'Anti-Litter Leagues' to organise litter picks and educate children on the litter problem. In 1934, CPRE's Sir Geoffrey Fry, former Private Secretary to Prime Minister Andrew Bonar Law, formed Wiltshire's Anti-Litter League; Gloucestershire's saw CPRE volunteers act as wardens on Minchinhampton Common, gathering signed anti-litter pledges from the public. The initiative proved so successful that it led to a nationwide Countryside Warden scheme and a 'Code of Conduct for the Countryside', created by CPRE with P.B. Nevill and Lady Armitage Smith of the Scouts and Guides respectively. Hundreds of wardens were soon enrolled across the country with their most important task being to 'suggest that rubbish can be carried home as easily as it was brought out'.

TIDYING UP FOR QUEEN AND COUNTRY

CPRE's Herbert Griffin made an anti-litter radio broadcast in 1937 to promote a nationwide drive for litter prosecutions, but it would not be until 1950 that litter regained its pre-war prominence during preparations

LITTER. For *thirty* years the C.P.R.E. and others have tried to check the litter nuisance by educating public opinion. Has not the time come to reinforce education by prosecutions under the byelaws and the imposition of suitable fines on those who indulge in this filthy and wasteful habit? See page 51.

for the following year's Festival of Britain. Anxious to divert public attention from the controversial South Bank Exhibition and Battersea Park 'Fun Fair', organisers enlisted CPRE's help to make the Festival a countrywide celebration. Herbert Griffin wrote to each parish council to instigate a 'tidying-up' campaign, and encouraged the Planning Minister Hugh Dalton to launch an anti-litter campaign to ensure Britain looked at her best for visitors. The campaign slogan was 'Keep Britain Tidy'.

In 1952, Elizabeth II's aunt, the Princess Royal, opened CPRE's National Conference in Harrogate by confessing that although 'paper bags and broken glass have much the same effect on me as a red rag to a bull', litter was 'something on which we can all help'. The Princess noted that 'next year many thousands of visitors will come to England for the Coronation. We want our fellow countrymen from the Dominions and Colonies to see England at her best and most smiling.' Griffin again wrote to parish councils, this time suggesting that they commemorate the

From CPRE's Annual Report of the Coronation year of 1953. The rising tide of rubbish convinced CPRE's Herbert Griffin to support the Women's Institutes' proposals for the creation of a Keep Britain Tidy Group the following year.

Coronation by providing litterbins on village greens, playing fields and bus stops.

Unfortunately, by 1954 Griffin had been forced to concede that 'the plague has, if anything, increased', but seconded the WI resolution to establish a 'Keep Britain Tidy Group' at their Westminster City Hall meeting the following year. At the group's first Annual Conference, Lord Attlee declared himself to be in favour of 'a few exemplary punishments for offenders with plenty of publicity attached to them', while the Queen Mother sent a message expressing her 'great pleasure of the growing interest in the anti-litter campaign'.

By 1958, the campaign had helped pass a Litter Act, which helped to promote 'litter consciousness' by introducing fines of up to £10 for littering. Throughout the 1960s, abandoned vehicles became a growing problem, together with discarded plastic packaging which people found difficult to re-use. By 1972, the Government made a fivefold increase in the annual budget for Keep Britain Tidy, allowing it to start educational publicity campaigns fronted by The Wombles. Despite raising awareness among a generation of children, the problem of litter continued to grow. The Keep Britain Tidy slogan was dropped in 1985, almost as an admission of defeat, while responsibility for litter was handed to the UK2000 consortium of business leaders chaired by Richard Branson.

In 1988, Mrs Thatcher confessed that a relaunch of new 'Tidy Britain' projects had been allocated just £60, and this had been spent on scattering litter around St James's Park after dark and then picking it up again later, for the benefit of the news cameras. That October, Lord Hayter asked the Government 'whether they have given up hope of persuading people to keep Britain tidy'. Measures suggested by the Tidy Britain Group did help the Environmental Protection Act 1990 to reduce litter by 13 per cent in its first year, but by 2001 Tidy Britain was absorbed by an awfully named new quango. Supposedly short for Environmental Campaigns, ENCAMS was intended to monitor indicators of declining 'local environmental quality'; counting litter was evidently easier than trying to stop it.

A NEW DETERMINATION TO 'STOP THE DROP'

In July 2007, Bill Bryson was elected President of CPRE, bringing to the role a personal and passionate commitment to campaign for a litter-free English countryside in the vacuum left by Keep Britain Tidy. April 2008 saw the launch of CPRE's new flagship campaign 'Stop the Drop', which aimed to challenge litter and fly-tipping and restore pride in local communities. During Bryson's five-year presidency, Stop the Drop made huge strides towards tackling an issue which now costs the country £1 billion a year to clear up. Official 2013 surveys showed that the percentage of sites with an unacceptable amount of litter had fallen to 15 per cent – down from 21 per cent before Stop the Drop was launched. While this cannot be solely attributed to the hard work of the local community groups inspired and equipped by the campaign, it reflects the benefits of a high-profile campaign with passionate leadership.

The launch of Stop the Drop might well have inspired the resurrection of the Keep Britain Tidy brand in 2009, but more tangible achievements include forcing Network Rail to clear 130 sites of litter and fly-tipping and inspiring over 600 new 'LitterAction' groups to collect 120,000 sacks of litter from their communities. In 2014, the campaign convinced the Government to provide local councils with additional powers to deal with litter thrown from vehicles, and was successful in lobbying for the introduction of a 5p charge on the plastic bags that so visibly blight the countryside.

Rural litter may sometimes be portrayed as a relatively trivial problem, but our willingness to put up with it (or determination to tackle it) says a lot about our pride in the countryside. It is notoriously difficult to change public behaviour, but with the bag charge seeing an immediate impact (Tesco reported a 78 per cent fall in plastic bag use in the first month of the scheme) and inspirational local activists encouraging sustainable lifestyles, there is real hope that we will be able to kick the litter habit.

ABOVE
Bill Bryson launches CPRE's Stop the Drop campaign with litter-picking stalwarts The Wombles.

OPPOSITE
F.H.K. Henrion's classic 1960s poster for Keep Britain Tidy forces us to ask whether we will ever leave the 'mess-age' behind.

IDEA 16 • ANTI-LITTER

161

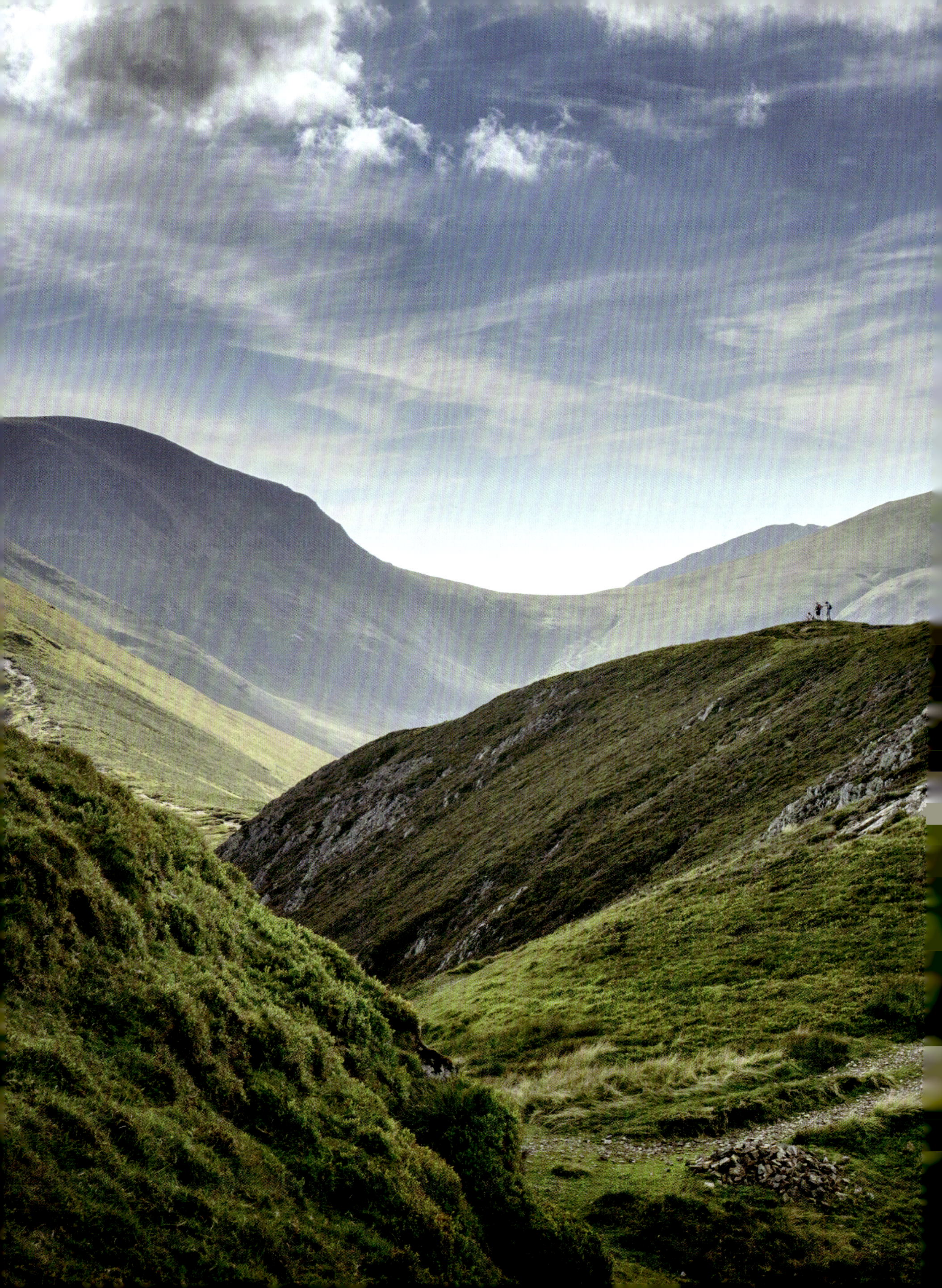

Idea 17

The Right to Roam

introduced by Julia Bradbury

The 1932 Mass Trespass of Kinder Scout in The Peak District was one of the most successful social uprisings of our times, when walkers and workers collided with gamekeepers and the landed gentry over access to the countryside. This fracas arguably led to the creation of our National Parks and the thirst for access encouraged generations of us to develop a special connection with the landscape. Since the Right to Roam became a legal reality in 2000, millions more of us have discovered places that have been out of reach for centuries. It's hard to imagine a time when it was ever difficult or dangerous ('Get orf moi land!') to explore our countryside, but the Right to Roam shouldn't be taken for granted in England. Though pioneering campaigners had been championing the cause of access for decades, it was the violent scenes of the Kinder Scout Trespass that became embedded in the national psyche as a potent reminder of the passions that can be stirred when people are denied free enjoyment of the beauty and tranquillity of their countryside. Walking evangelists like Alfred Wainwright and associations like the Ramblers' have fuelled a national love of exploring our landscape. The millions who step out onto England's footpaths and national trails every week are proof that our Right to Roam remains a precious and deeply appreciated gift.

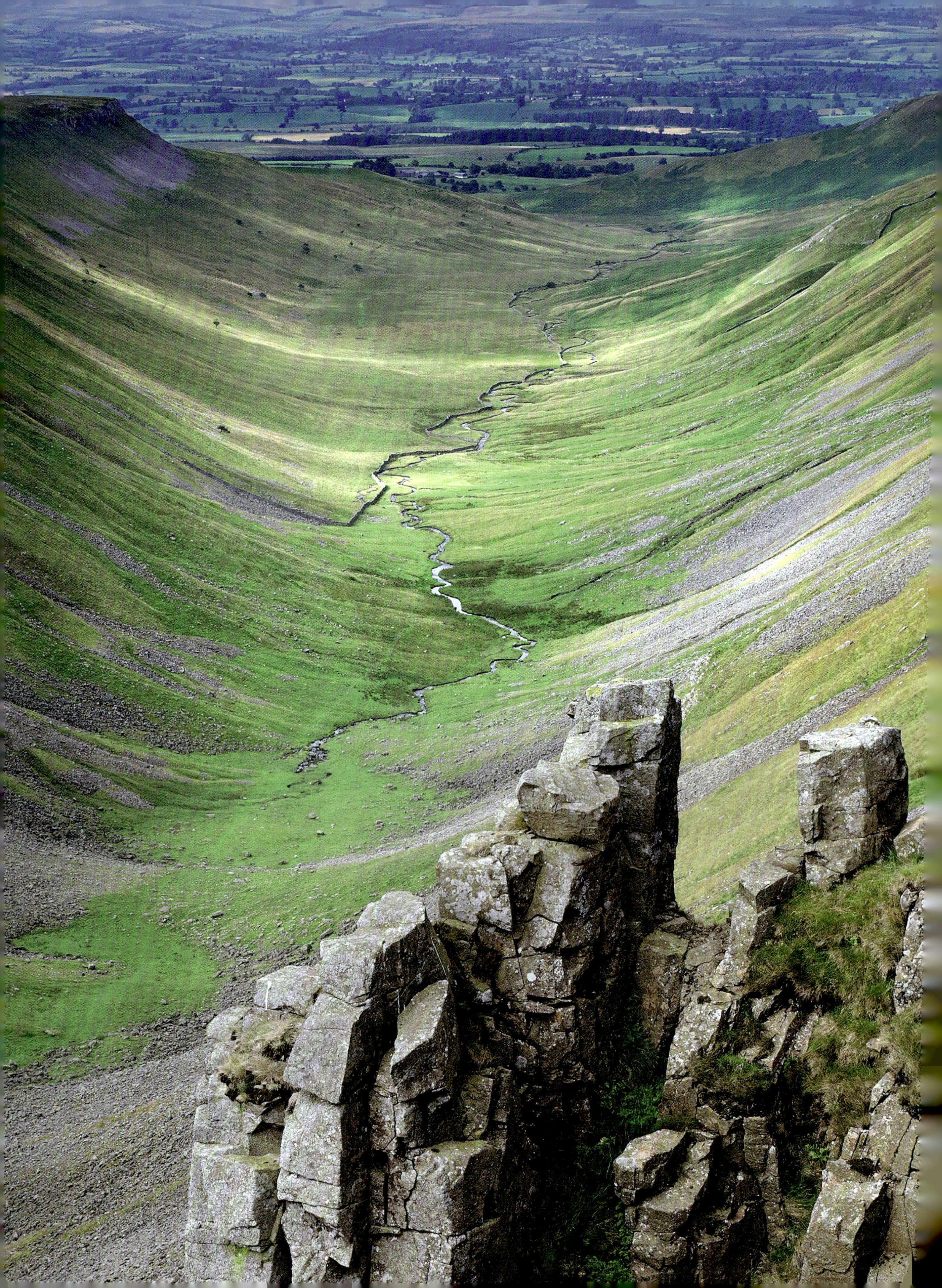

PAGES 162–3
Causey Pike in the Lakeland Fells; a classic walking route beloved of ramblers from Wordsworth to Wainwright.

OPPOSITE
The view of High Cup Gill from the Pennine Way National Trail, Britain's first long distance footpath and the inspiration of the Ramblers' campaigner Tom Stephenson.

The common right to walk across the countryside was taken completely for granted until the Normans, who, despite introducing the concept of property rights, were still obliged to maintain paths between villages and hamlets. These ancient paths became enshrined in law by an Act of 1555 forcing every parish to keep them clear. By the early nineteenth century, large-scale enclosure had already begun to curtail freedom of movement in the countryside, a loss documented by the poet John Clare:

> Crossberry Way and old Round Oak's narrow lane
> With its hollow trees like pulpits, I shall never see again

William Wordsworth was one of the first to challenge the wilful obstruction of public rights of way – an increasingly common tactic of landowners trying to keep walkers from disturbing their grouse moors. The practice was made illegal by the Highways Act of 1835, but Wordsworth found himself accused of trespassing on an old green lane the following year. His companion, a nephew of Samuel Taylor Coleridge, observed that 'he evidently had a pleasure in vindicating these rights and seemed to think it a duty'. Some versions of the story have Wordsworth physically tearing down a wall blocking the ancient path, something Samuel Plimsoll and his followers were equally prepared to do fifty years later. While protesting the blocking off of the summit of Latrigg in the Lake District, the radical MP and inventor of the Plimsoll Line joined 2,000 walkers armed with spades and pick-axes to 'persuade' the landowner to reopen the path in perpetuity.

A THIRST FOR ACCESS

Plimsoll's success was a rarity. A 12,000-strong 1896 protest in the Bolton moors failed to keep Winter Hill open to walkers, as ringleaders were fined heavily and the law asserted the rights of individual landowners over the freedom of the public. In *This Land Is Our Land*, Marion Shoard argues that the First World War gave the fledgling rambling movement a new sense of entitlement to the land they had fought for – land that now offered free recreation during the prolonged period of mass unemployment in the 1920s. The most obvious way to secure access was to acquire land and donate it to the National Trust, and in 1927, the CPRE Sheffield and Peak District branch acquired Longshaw, a 747-acre estate belonging to the Duke of Rutland. The land was originally offered as building plots on the open market, but with the help of a public appeal and a sizable overdraft, campaigner Ethel Haythornthwaite was able to hand over the land to G.M. Trevelyan of the National Trust.

Buoyed by the success, Trevelyan used CPRE's 1930 conference on the 'Use and Enjoyment of the Countryside' to encourage 'our town population and young people, to go out into the country and use it properly in the way of walking and camping'. Then, in that year's debates on the Rural Amenities Bill, the novelist and MP John Buchan made a passionate defence of free access to footpaths:

> England would be a dull place and a poor place without those footpaths which wander about in woods and meadows. I would like to see an Act of Parliament passed to schedule existing footpaths and perpetuate them, to create a new Domesday Book, because I believe that that would give people a sense of responsibility for rural beauty, since they would feel that they were partly the owners of it.

The following year, Sir Lawrence Chubb of the Commons and Footpaths Preservation Society worked with CPRE and the Ramblers' to encourage landowners to to open up their land to the public, even suggesting incentives like tax concessions. CPRE helped ensure that the Rights of Way Act 1932 contained powers to make established paths across private land 'public rights of way', while the Ramblers' Federations were beginning to mobilise up to 5,000 walkers for their rallies. Together with the British Workers' Sports Federation they campaigned against the inaccessibility of the Peak District – the favoured destination of the workers of Manchester and Sheffield. The Kinder Scout mass trespass of 1932 asserted, in the words of their leader – the twenty-year-old Mancunian mechanic Benny Rothman – 'the rights of ordinary people to walk on land stolen from them in earlier times'. The violent clash between walkers and gamekeepers was soon immortalised by the folk singer Ewan MacColl's 'Manchester Rambler':

> I'm a rambler, I'm a rambler from Manchester Way,
> I get all me pleasure the hard moorland way,
> I may be a wage slave on Monday
> But I am a free man on Sunday.
>
> [The gamekeeper] said 'All this land is my master's'
> At that I stood shaking my head
> No man has the right to own mountains
> Any more than the deep ocean bed.

RAMBLING BY AGREEMENT

At the time of Kinder Scout, CPRE commissioned a detailed report on 'Access to the Moorlands in the Peak District' by Phil Barnes, a renowned climber and photographer. Barnes showed why Kinder Scout had been such a flash point, arguing that its 'cloughs and edges afforded unquestionably the best scenery in the peak'. He felt the frustrations of the trespassers at the

IDEA 17 • THE RIGHT TO ROAM

Easter Day walkers look forward to the view from the top of Streatley Hill in Berkshire, just weeks before the Kinder Scout trespass of 1932. The Times reported 'special "hikers" trains' leaving Paddington with 'large numbers of city people, delighted by the prospect of a change from the winter routine'.

. . . serious lack of footpaths in the moorland areas of the Peak District: 'The Downfall' is the best waterfall in the District and undoubtedly one of the most impressive in the country, but can only be seen from considerable distance.

Barnes argued that because the 'trespass routes' were the most direct way to the beauty spots, re-opening them would actually reduce damage to game bird habitats on the rest of the moors. He also felt increasing access would help encourage the growth of the youth hostel movement, recently endorsed by the Prince of Wales' (soon to be Edward VIII) opening of Derwent Hall in 1932, a building which CPRE helped to secure for the Youth Hostel Association.

Soon after the Barnes report was published, Sir Lawrence Chubb gave a speech to CPRE's annual conference – specially convened in Buxton – calling on landowners to 'remove a deep-rooted feeling of discontent' by allowing townsfolk to 'learn to know rural England'.

Barnes' recommendation for CPRE volunteers to act as Countryside Wardens was put into action by 1935, with *The Times* hoping it would encourage landowners to throw more land open to the public as people would be on hand to promote responsible behaviour and stop the sort of vandalism 'which often caused facilities for access and privileges hitherto granted to the public to be withdrawn'.

ACCESS AS A RIGHT

Despite some progress on permissive access, those seeking the right – rather than permission – to explore the countryside felt let down by the landowners. The Access to Mountains Act of 1939 – seen as the Establishment's response to Kinder Scout – was, according to campaigner Tom Stephenson, 'not a ramblers' charter, but a landowners' protection bill'. Conservative pressure had even made 'trespass' a criminal offence for the first time, prompting H.G. Wells to write an update of Thomas Paine's *The Rights of Man*. Ranking access to the countryside alongside nourishment, housing, healthcare and employment, Wells called for 'the right to roam over any kind of country, moorland, mountain, farm, great garden or what not'.

If the 'right to roam' seemed closer after 1945, it was in no small part due to the representations of the Ramblers, whose first Secretary, Tom Stephenson, was instrumental in convincing the Government to incorporate long-neglected access provisions into a Bill ostensibly

A snow-covered Kinder Scout gives one of the most majestic views of the Peak District. The anger of the 1932 trespass had been bubbling up since the Enclosure Acts of Wordsworth's era allowed landowners to block off ancient packhorse routes like Alport Castles Bridleway and The Duke of Norfolk's Road.

IDEA 17 • THE RIGHT TO ROAM

THE RIGHT TO ROAM
THE RAMBLER AND THE COUNTRYSIDE

Price Two Shillings

Inspired by H.G. Wells' slogan, the Ramblers' Association has had many successes in increasing access to the countryside.

designed to create National Parks. On introducing the National Parks and Access to the Countryside Bill, Lewis Silkin was clear that without it, 'everyone who loves to get out into the open air and enjoy the countryside' is 'fettered, deprived of their powers of access and facilities needed to make holidays enjoyable'.

The eventual Act of 1949 could not come too soon for CPRE, which was so concerned about the disrepair of ancient paths that it felt 'that, at the very time when more and more people are turning to country walking as a health-giving and enjoyable recreation, the intricate system of country footpaths, which has hitherto been one of the most valuable and distinctive characteristics of the English countryside, is in imminent danger of breaking down'.

The Act required every county council to compile a 'definitive map and statement' of all public rights of way, but most failed to take practical steps to open them up, perhaps unsurprisingly given the representation of landowners on local councils at the time; Stephenson confessed that he always felt 'that the rights of the rambler were a featherweight against the rights of property'.

More bad news was averted in 1966 when the Ramblers defeated Government plans for a significant reduction of the rights of way network. The Labour peer and future CPRE Chairman Lord Kennet argued that the Government should be aiming to improve access; working with his CPRE predecessor Lord Molson, he helped ensure the subsequent Countryside Act 1968 greatly widened the scope of the 1949 Access provisions by opening up woodlands, lakesides and river banks.

OPEN FOR PLEASURE

But by the 1980s, the rise of intensive farming had seen many paths ploughed up or fenced off; the Ramblers helped halt the trend by winning greater protection for paths on farmland through the Rights of Way Act 1990. By 1997, their campaigning secured a Labour manifesto commitment to a 'Right to Roam', which Prime Minister John Major promptly declared would be tantamount to a 'charter for rural crime'. When the Countryside and Rights of Way Bill was introduced in 2000, Environment Minister Michael Meacher said it 'finally brings to reality the dream of Lloyd George that nobody should be a trespasser in the land of their birth'.

Taking another four years to create maps showing exactly where walkers could enjoy new open access land, the Right to Roam finally came into force in England on 31 October 2005. Almost 11,000 square miles of mountain, moor, heath and down have since been opened up, allowing people to freely stray off the footpath and explore the deepest reaches of the countryside. There is still plenty of room for improvement, and Marion Shoard has pointed out that the Act still leaves 90 per cent of England and Wales out of bounds. However, there is no doubt that access to areas like the Yorkshire Dales has been transformed. Before 2000, just 4 per cent of the Dales was 'open for pleasure'; now walkers can enjoy unfettered access to almost two thirds of it. Successes like that mean the logical next step on the path could be for England to follow Scotland in granting the right to universal access to the land.

ABOVE
The path to Mam Tor is one of the most popular walking routes in the Peak District.

OPPOSITE
The view towards Smearsett Scar from Goat Scar Lane. The Right to Roam has opened up many more walking routes and stunning views in the Yorkshire Dales since 2005.

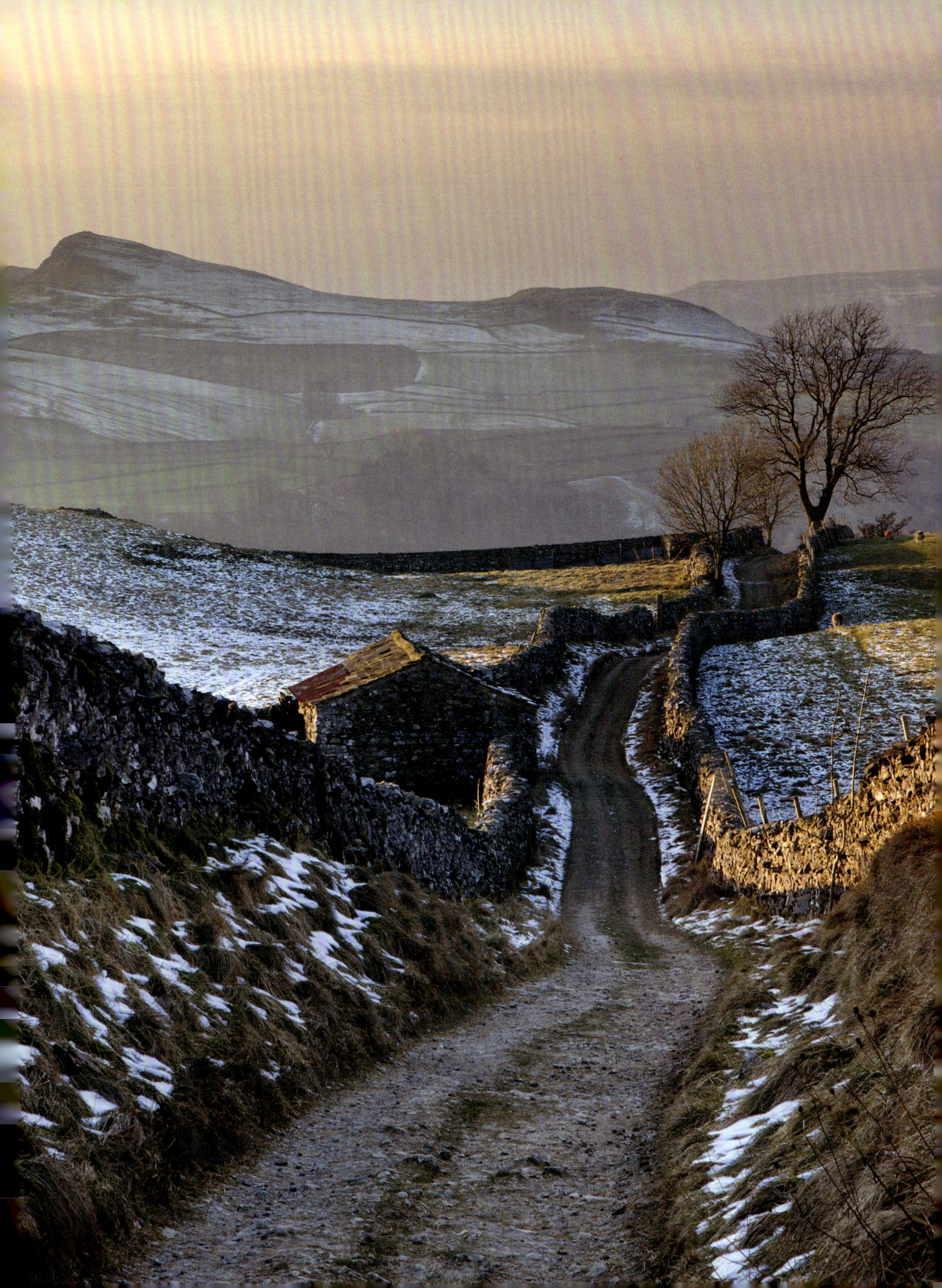

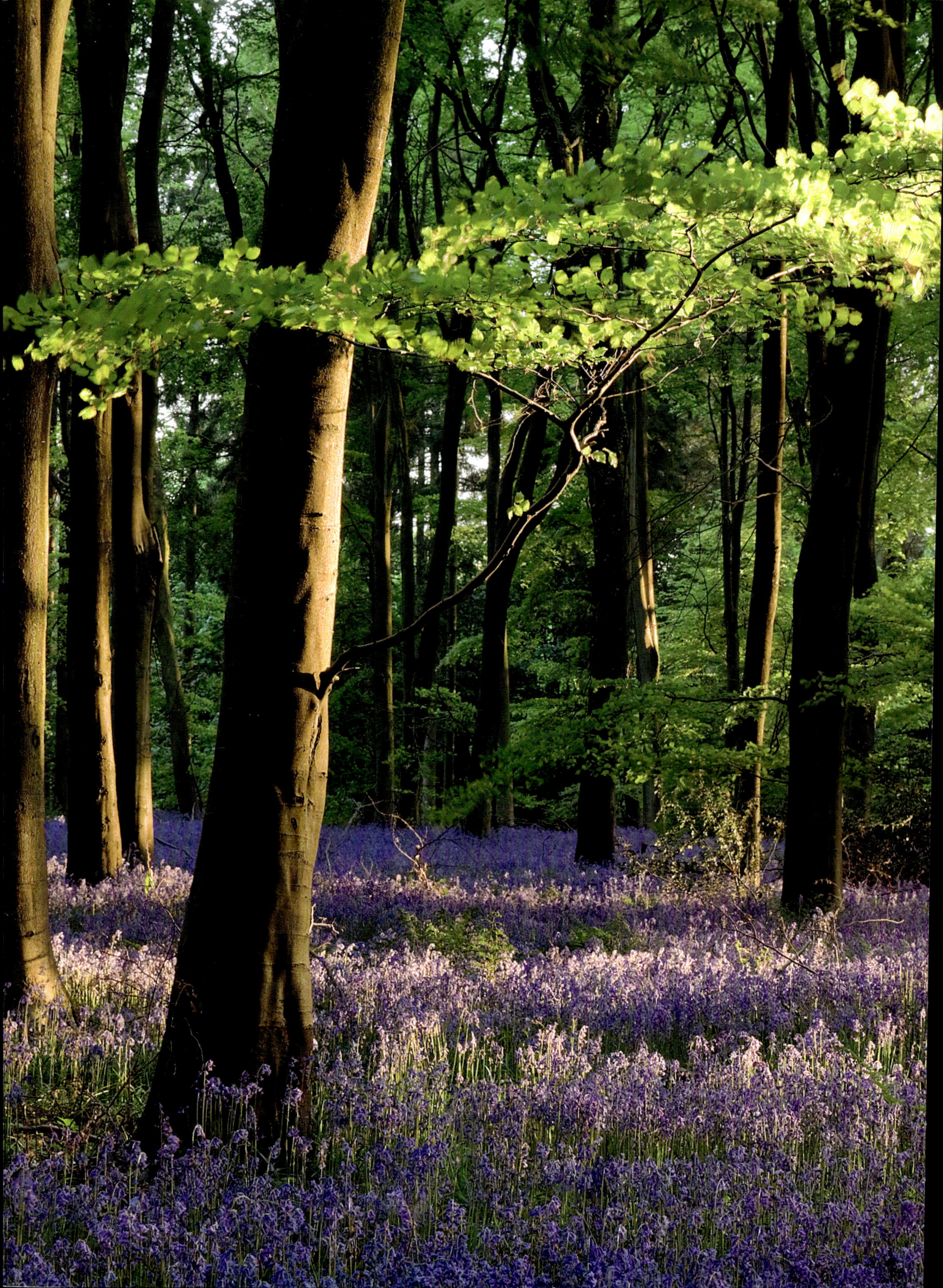

Idea 18

Saving our Forests

introduced by Jeanette Winterson

Woods are not a luxury; if we want to re-stabilise our climate and our planet, and restore the beauty of our landscape the re-wilding, we will need to begin by planting trees. From John Evelyn to John Ruskin, the defenders of our woodland always knew that replenishing our native trees was vital part of our stewardship of the countryside. Following in the fine traditions these pioneers established, many of us are becoming both more aware and more protective of our woods. The Government's ugly proposal to sell off much remaining forest to commercial interests aroused unusual passion. It wasn't only our walks and weekends that were at stake – it was our imagination.

Our love of the forest is, I think, a profound love that is not nostalgia but a living memory of pre-industrial life. The Industrial Revolution and our mass migration into the cities only began little more than two hundred years ago – not much time at all in the life of a great oak. With so many positive initiatives to increase England's tree-cover, we must not forget our increasingly vulnerable ancient woods. If we can learn the lessons of history, future generations will be able to go into the forest and lean on an ancient tree, feel the past at their back, and see green shoots unfurl in front of them.

History shows us that human interaction was generally bad news for trees until the turn of the twentieth century. The Domesday Book recorded that only around 15 per cent of England was covered by woodland; five millennia of farming clearances and harvesting for fuel and building materials had reduced that figure from 75 per cent. By the end of the nineteenth-century woodland cover had dropped to below 5 per cent of England's countryside, but had recovered to 8.4 per cent – equivalent to over 4,250 square miles – at the beginning of the twenty-first.

So, how was the demise of our woodlands halted and reversed? A major reason has been the work of the Forestry Commission, created by the 1919 Forestry Act in response to First World War timber shortages. Somewhat ironically, the original inspiration for a national tree-planting scheme came from a German, Wilhelm Schlich, who came to England in 1885 to become the University of Oxford's Professor of Forestry. By 1887, Schlich's teaching had made it a source of national shame that England's forests were 'treated with a rude recklessness which would astonish and disgust a German forester'. Ahead of a State visit that year, *The Times* felt that 'we should be more afraid of the judgement Prince Bismarck might pass on British woods than on British politics and military administration'. Prime Minister Gladstone was so fond of using tree-felling exhibitions as displays of industrial progress that Bismarck presented him with a young oak to take home and plant.

CULTIVATING ENGLAND'S LOVE OF TREES

John Evelyn's *Sylva* of 1664 is still hugely influential as the first treatise on sustainability (we need to plant at least as many trees as we cut down), but at the time it created a kind of mania for tree-planting. In the early eighteenth-century, the poet Joseph Addison ventured that 'the love of woods seems to be a passion implanted in our natures', while William Gilpin considered that the picturesque was nothing without trees – 'the grandest and most beautiful of all the productions of the earth'. By the mid-nineteenth century, Cowper, Clare and Wordsworth had all lamented the felling of trees in verse, and in 1884, Ruskin avowed that it should be forbidden 'to cut down for mere profit, trees whose loss would spoil a landscape'.

Despite the growing affection for trees in England, the decline of England's woodland cover reached crisis point by the First World War. The remit of the fledgling Forestry Commission was to rebuild and maintain a strategic timber reserve, using a good deal of freedom to acquire and plant land. With agriculture in depression, they were able to buy 600,000 acres in their first decade, but concerns raised by CPRE about the landscape impacts of blankets of conifers led to the planting of native hardwoods alongside roads, and greater attempts to blend the new plantations into the countryside.

In April 1936, CPRE's President Lord Crawford told the Lords that 'we have been pressing the Forestry Commissioners to form in the Lake District

PAGES 172–3
Micheldever Wood in Hampshire typifies the Forestry Commission's increasing role in promoting the use of the nation's forests for recreation.

what we call a "sacrosanct area" in which no afforestation can take place without our consent'. Crawford also pointed out that because the Commission had to acquire whole estates in which some land was not suitable for planting, the public would gain 'an area of 5,000 acres over which they will be free to roam unembarrassed in any way by afforestation'.

THE CONIFER QUESTION

Though Crawford personally admired their 'wonderful colouring', the Commission's choice of fast-growing foreign conifers was not to everyone's taste. G.M. Trevelyan objected to the creation of 'German pine forests', while Dudley Stamp later observed the violent opposition to England being 'reclaimed' by non-native trees. Madeleine Bunting recently suggested 'the coniferous plantations were an expression of the anxieties and fears' brought on by the wars:

> Twice within twenty-five years Britain . . . had found itself humiliatingly short of a humble commodity like wood. It was these crises that provided the Commission with its sense of urgency and, even more importantly, the legitimacy to clear and reorder the land.

In July 1936, a joint agreement was signed between CPRE and the Forestry Commission to prevent any such 'reordering' in the 300 square miles of the Lake District's central fells. Trevelyan expressed delight that the meadows near Esthwaite Water 'will neither be planted nor built upon, but will remain as the schoolboy Wordsworth knew them'. CPRE's negotiations had been greatly assisted by the Friends of the Lake District's 'red line' maps, drawn up as a forest-free 'exclusion zone' by John Dower and later used as the basis of the National Park boundary.

The Commission then began an experiment in promoting public access – with basic footpaths and shelters – on their 400,000 acres of 'unplantable' land. Official 'National Forest Parks' were soon created in Wales and Scotland, with the Forest of Dean set to be the first in England; socialist MPs like Thomas Johnston were delighted, if not a little bemused, that the public would be allowed to 'see the beauties and the glories of a nationalised land system which a Conservative Government are happily inaugurating'.

The Second World War and its aftermath meant National Forest Parks were put on hold and the emphasis was put on replenishing the exhausted stock of timber; instead of becoming parks for enjoying the 'beauties and glories' the Forest of Dean and the New Forest bore the brunt of wartime felling, with almost all conifers over twenty years old cut down. Thanks to CPRE campaigning, the eventual replanting included more broadleaves and

Edward McKnight Kauffer's atmospheric poster design of 1932 combines with lines from Milton to sum up England's spiritual need for trees.

was increasingly planned to avoid good quality farmland. The 1951 Forestry Act even allowed the Commission to make greater concessions to 'amenity' in new planting. Having published *Forestry in the Landscape* in 1966, the landscape architect Dame Sylvia Crowe was a logical choice to become the Commission's first landscape consultant, prioritising the planting of native species and the blending in of plantations with surrounding countryside.

From 1967 the Commission's aims were enlarged to encourage recreation, and by 1970 they were receiving 15 million annual visits to their 3 million acres of land – two thirds of which was under tree cover or scheduled for planting. Historian Victor Bonham-Carter noted how the Commission's increasingly sensitive management had encouraged wildlife: 'at Grizedale in Westmorland for instance, the greylag geese have come back, accompanied by the mallard and tufted duck.' A hundred million broadleaved trees were planted between 1980 and 2000, often around the edges of plantations to hide the straight rows of conifers. In 1988, CPRE led a campaign that convinced Chancellor Nigel Lawson to scrap tax incentives favouring blanket conifer plantations in upland landscapes.

A TRANSFORMATION IN TREE PLANTING

The Commission planted mostly native trees after 1990, often through their initiative to create ten Community Forests on 25,000 acres of underused land near England's main cities. Said to be, collectively, the largest regeneration scheme in England, this resulted in half of England's population now living

The view from Loughrigg is an example of how the Lake District fells have largely avoided insensitive afforestation.

within easy reach of a Community Forest. In the mid-1990s, The National Forest was created to regenerate 200 square miles of disused mining landscapes in the East Midlands; over 8 million trees have been planted so far. Thanks to these initiatives, the proportion of England covered by woodland has doubled from 5 per cent in 1945 to 10 per cent today. And this is despite the loss of 30 per cent of our elms to Dutch elm disease in the 1970s, and the 15 million trees that were blown down during the 1987 gales.

The recent work of the Forestry Commission has enhanced landscapes, created new habitats, and brought millions of people closer to nature – all at a total cost of 30p a year for each taxpayer. And yet, in 2011, so much of this progress was put at risk when the Coalition Government attempted to sell off the forests to private bodies. With the Forest of Dean particularly vulnerable to access changes, local CPRE campaigner Colin Evers articulated the thoughts of thousands: 'For hundreds of years it was used by monarchs to hunt in, but today it is one of our forests, a national gem, managed for all of us by the Forestry Commission.' Over 530,000 people signed an online petition to keep the 637,000 acres of the Public Forest Estate in public hands.

In the end, the Government was forced by public opinion and pressure from the Save Our Forests campaign to perform a screeching U-turn. Before the dust had settled Ash Dieback disease struck, threatening one of the most iconic trees of the English landscape. CPRE called on the Government to use the UK's Contingency Fund to tackle the disease, with their former President Jonathan Dimbleby saying 'ash is in so many copses, woods and hedges – we need to encourage more planting of trees'. The Government responded by setting out plans to invest £6 million in planting 4 million trees in England in 2014–15, creating 5,000 acres of new woodland.

The plans are further evidence of the remarkable evolution of the Forestry Commission and the growing political recognition that forests should be an integral part of the countryside. For the public, of course, this had never been in doubt; the Tree Council are doing wonderful work in mobilising 250,000 people to plant a million trees in their annual *National Tree Week*, and the Woodland Trust helped communities plant 6 million trees in sixty Jubilee Woods in 2012. With the Forestry Commission's 'Big Tree Plant' successfully planting 1 million trees in urban and rural communities by 2015, there is real hope that England could get close to the 15 per cent woodland cover recorded by the Domesday Book.

A 1960s Forestry Commission poster warning against forest fires; where the Commission was itself once accused by some of devastating the countryside with excessive conifer plantations, by 2011, the environmentalist Jonathan Porritt could argue that its woodlands had 'a higher percentage of its Sites of Special Scientific Interest in 'good condition' than the RSPB.

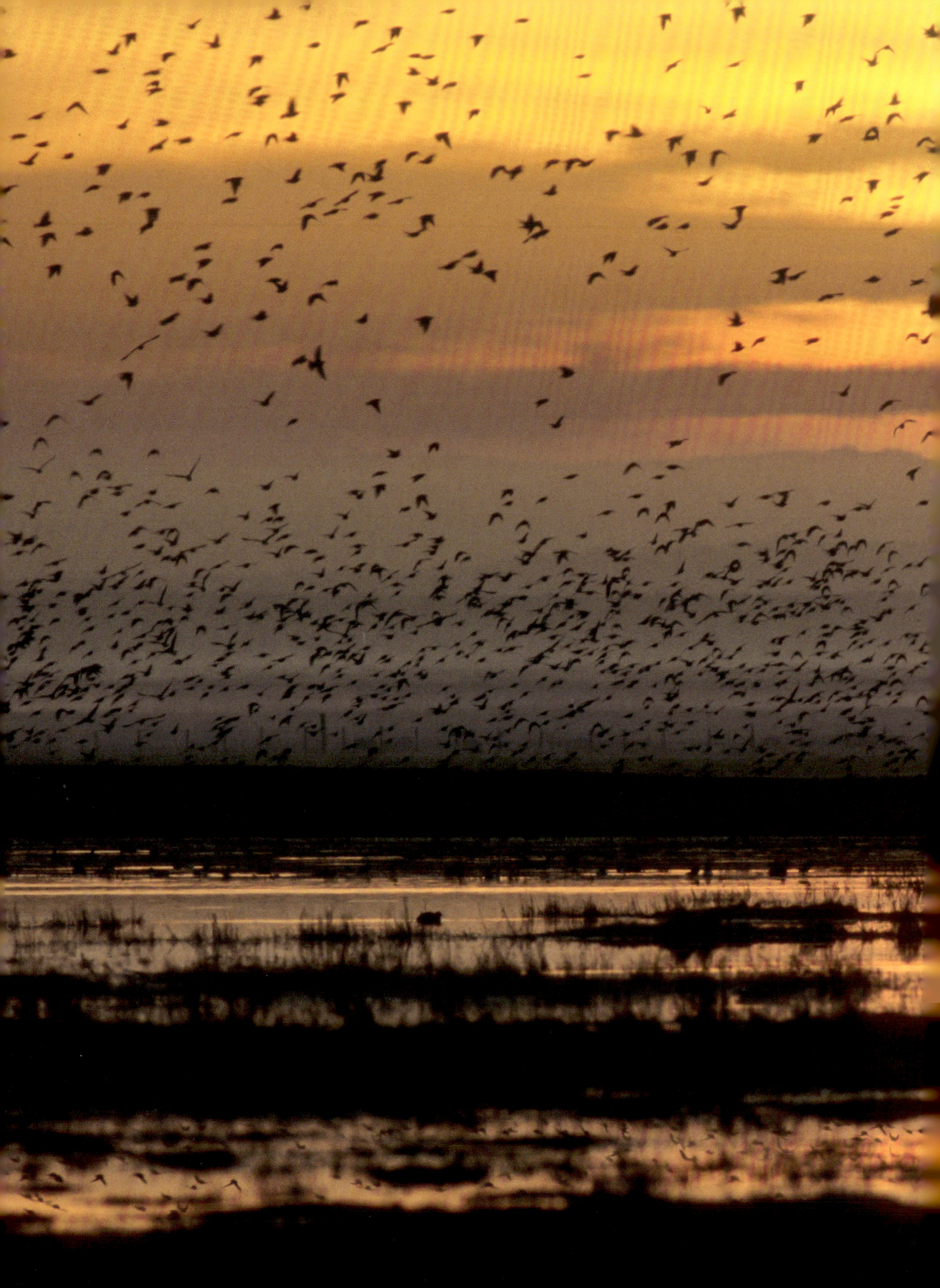

Idea 19

Nature Reserves

introduced by Virginia McKenna

At last! The concept that the countryside should be 'one great nature reserve' must be the way to follow. The alternative is too fearful to contemplate.

As things appear to be going – sometimes obviously, sometimes so subtly that we hardly notice that woods and fields have vanished, and the forest we are seeing is of bricks and mortar – that alternative is almost upon us.

All of us understand that when time began the creatures that walked on the earth, swam in the seas and rivers or flew in the sky, created the environment. When human populations were small and scattered this was sustainable. But now it is all breaking down. Even Areas of Outstanding Natural Beauty are vulnerable.

Where I live in Surrey, an oil company wants to prospect for oil on Leith Hill. The area is an AONB, of great natural beauty and abounding with wildlife. It has wonderful forests and hedgerows, and is a place where we can find peace and calm in our turbulent world. It is a small example of what our natural environment faces, as material considerations overturn more sensitive, emotional and profound ones.

What a bereft and poor place the world will be if nature is undermined, undervalued and is wrecked even further by machines and chemicals. Let us pay more attention to Waterton – not perhaps as eccentric as he appeared in 1817. Individuals have always inspired us to think differently and open our eyes to various aspects of life – as you will read in the timely wake-up call of this chapter. This is not 'crying wolf' – it is a warning that unless we open our eyes and see and feel the value of the world and its creatures, one day we will see only buildings and concrete and ourselves, only hear machines.

Not a world I would ever wish to live in or wish on anyone else.

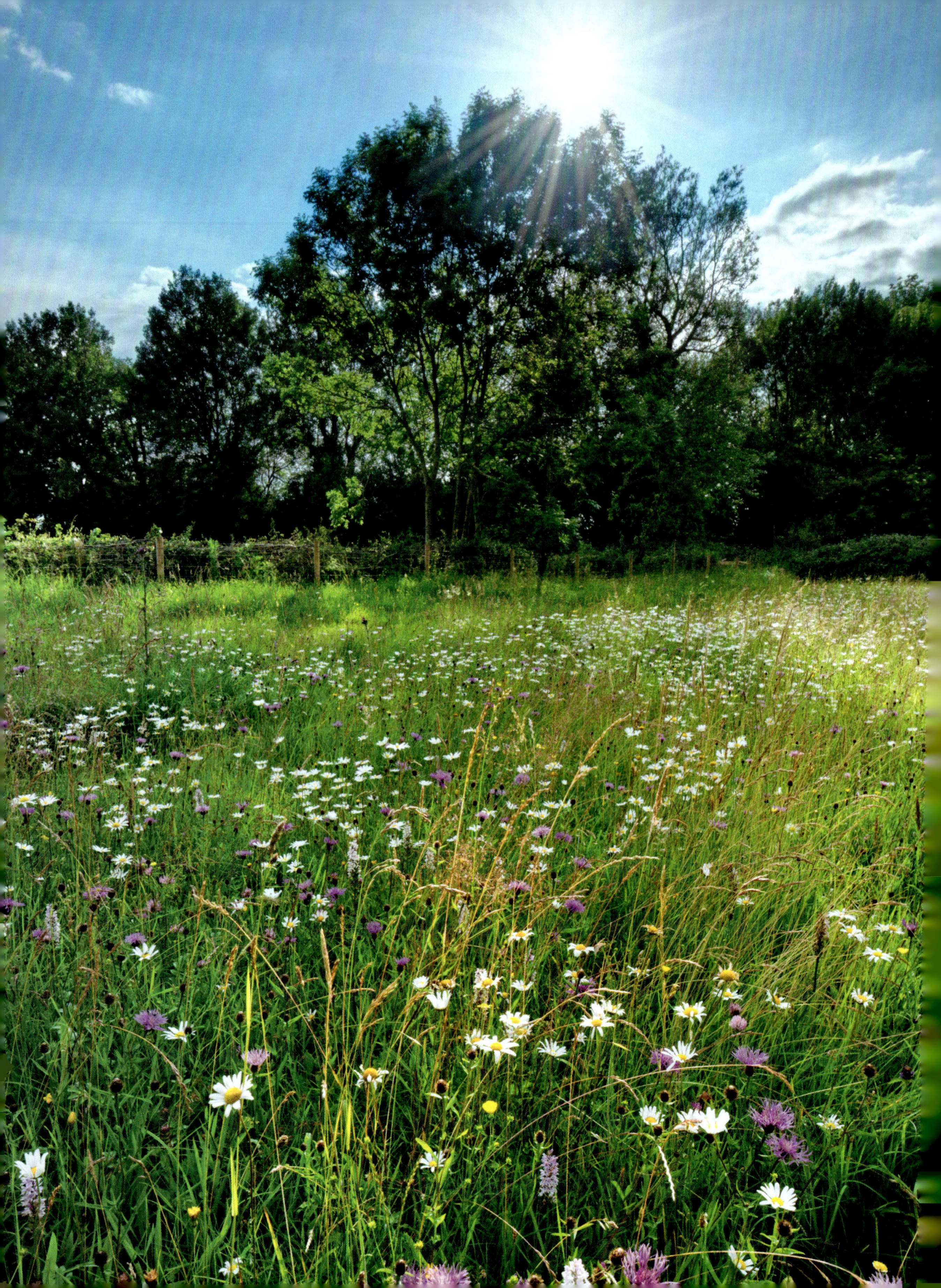

PAGES 178–9
Elmley National Nature Reserve in Kent is Southern England's largest area of coastal grazing marsh; as well being as a haven for wildlife, it provides a source of inspirational beauty and tranquillity for human visitors.

OPPOSITE
Clattinger Farm only became a Nature Reserve when it was acquired by the Wiltshire Wildlife Trust in 1996, but is a symbol of the positive work being done to save rare grasslands and the species that rely on them.

Despite the headline-grabbing claims that back gardens and derelict sites are the perfect places for wildlife, the countryside is still the largest and most diverse habitat we have. Unfortunately, it is true that while animals are no longer hunted to extinction, their habitats have suffered from nearly two centuries of pollution and urban sprawl, and over seventy years of intensive farming.

Even before the destruction of habitat was linked to wildlife loss, one pioneer had an inkling that human interference could threaten the survival of whole species. The idea of a reserving space for nature was hit upon in 1817 by the Yorkshire naturalist and explorer Charles Waterton, an archetypal English eccentric; on his expeditions to South America he would walk barefoot through the jungle hoping to attract blood-sucking specimens, even sleeping with a toe extending from his tent to entice passing vampire bats.

Determined to stop people shooting birds on his estate, Waterton began erecting a three-mile-long perimeter wall, sealing Walton Hall's lake and woodland off from human interference. By 1826, Waterton was effectively running the world's first nature reserve, allowing local people supervised access to witness the thriving wildlife. According to David Attenborough, Waterton was 'one of the first people anywhere to recognise not only that the natural world was of great importance but that it needed protection as humanity made more and more demands on it'.

Waterton's groundbreaking conservation work would not be emulated until Charles Rothschild decided to branch out from his day job at the family bank in London. As well as being a banker, Rothschild was the world's leading flea expert, assembling a collection of more than a million (now residing in the Natural History Museum). His love of insect collecting turned to alarm at the realisation that 99 per cent of England's fenland had been lost to the plough by the end of the nineteenth century. In 1899, he purchased Wicken Fen near Ely for the National Trust, and later made the nearby Woodwalton Fen his personal nature reserve. Rothschild understood that the 'good spots' for wildlife – the places where he found it easiest to collect specimens – needed protection.

SAVING HABITATS ON A NATIONAL SCALE

Henry Willett's purchase, for observation, of 5 acres of woodland bog near Abingdon in Berkshire in 1901 heralded a shift from collecting natural history to the scientific study of it. In 1912, Rothschild created a national Society for the Promotion of Nature Reserves (SPNR), producing a list of England's 182 best sites for wildlife by 1915. The First World War and Rothschild's death in 1923 meant the momentum was lost. In the meantime, Thomas Hardy's *Domicilium* of 1916 illustrated how much wildlife the countryside had already lost. Hardy recalled being taken for a country walk by his grandmother as a child and asking 'How looked the spot when first she settled here.' Her reply, 'Fifty years have passed since then,' dates her subsequent recollections to 1800:

>   . . . Snakes and efts [juvenile newts]
>   Swarmed in the summer days, and nightly bats
>   Would fly about our bedrooms. Heathcroppers [wild ponies]
>   Lived on the hills, and were our only friends;
>   So wild it was when we first settled here.

An autumn sunrise over the marsh landscape of Elmley National Nature Reserve on the Isle of Sheppey. Said to be Charles Dickens' inspiration for the setting of *Great Expectations*, Elmley is noted for its breeding waders, water voles and lizards, among many other species.

While only one of Rothschild's sites – Cley Marshes in Norfolk – was secured before 1945, Arthur Tansley's invention of the concept of the 'ecosytem' in 1935 helped promote the idea that protecting the countryside would help save species. By 1942, the Government's Scott Report on rural land use recommended the creation of State-run Nature Reserves, and Tansley's *Spectator* article the following year argued that they were essential for 'the preservation of our natural fauna and flora'.

By 1945, the Government's Wild Life Conservation Committee chaired by Tansley proposed seventy-three National Nature Reserves covering 70,000 acres of England and Wales, leading to a Royal Charter creating the Nature Conservancy to 'establish, maintain and manage Nature Reserves'. One of the earliest Nature Reserves established by the new body was Tansley's own favourite landscape – the view from Kingley Vale to Chichester – proving his argument that beauty and biodiversity could go hand in hand. But Tansley also knew that nature could thrive in the most unlikely places, reserving Minsmere Level in Suffolk despite, or rather, because of the fact it had been used as a bombing range during the war. For Tansley, this 'rough treatment' had made Minsmere 'exceedingly attractive to birds'.

THE RISE OF THE WILDLIFE TRUSTS

The creation of the Lincolnshire Wildlife Trust in 1948 had prevented the transformation of Gibraltar Point into a caravan site, and by the mid-1960s, the Nature Conservancy had acquired more than a hundred National Nature Reserves before losing its Royal Charter – and most of its funding – in 1965. Fittingly, Rothschild's SPNR had already begun to fill the vacuum, ensuring that almost every county had formed their own wildlife trust by the end of the decade. By 1981 the SPNR had gained its own Royal Charter, being renamed the Wildlife Trusts in 1995 and growing to a movement of over 800,000 members by 2011.

There are now over 220 National Nature Reserves totalling more than 233,000 acres, or approximately 0.6 per cent of England's countryside. Managed either by the Wildlife Trusts, RSPB, National Trust or the Government agency Natural England, they protect the most biodiverse parts of the landscape as a haven for countless rare and nationally important species. All but 21 of the 182 sites on Rothschild's original list are now designated Nature Reserves or Sites of Special Scientific Interest. The losses are a stark reminder of the fragility of unprotected landscapes, while the fate of the Wildlife Trusts' first historic reserve shows the importance of seeing Nature Reserves as part of wider conservation efforts. Charles Rothschild had thought he could protect Woodwalton Fen by buying it. Little did he know that the draining of the surrounding farmland would dry out the natural peat of the fen, meaning Woodwalton sank by an astonishing 13 feet while its wildlife vanished.

Thankfully, Woodwalton is being rewilded as part of the Great Fen Project to return drained land back to wild fen; it will be linked to the Holme Fen reserve by turning over 9,000 acres of Cambridgeshire prairie into a 'mosaic of attractive, species-rich and publicly accessible fenland, grassland, reedbed and woodland'. Despite this ambitious scheme, the RSPB's 2013 'State of Nature' report revealed that, while otters – and even beavers – are being seen in our rivers once again and red kites continue to thrive following their reintroduction, the majority of 'at risk' species are continuing to decline. This has prompted the RSPB to launch their Futurescapes campaign in the recognition that while 'nature reserves and protected areas were a good start . . . we can't just look after small pockets of land because nature is everywhere'. This refocusing reminds us that the ultimate goal of Rothschild and Tansley would surely have been to make 'nature reserves' redundant, by making the whole of the countryside a home for wildlife. The Wildlife Trusts are right that while biodiversity may have 'retreated into these last strongholds', Nature Reserves are still vitally important for preserving the species we hope will one day 'disperse and recolonise' the rest of the countryside. The acceleration of intensive farming – and especially the rise of chemicals being poured onto the land – after the Second World War means that aim is still far out of reach.

ABOVE
Intensive farming meant the destruction of priceless, ancient hedgerow habitats was a sadly common site in the 1970s and beyond.

OPPOSITE
With environmentalists increasingly thinking about landscape-scale conservation, schemes to leave field margins for wildlife are part of the process needed to make farming compatible with biodiversity once more.

THE COUNTRYSIDE AS ONE GREAT NATURE RESERVE

Rachel Carson's 1962 account of the impact of pesticides on wildlife, *Silent Spring*, helped ban the most damaging chemicals, but her role as a catalyst for the wider environment movement could be her greatest legacy. The formation of Greenpeace, Friends of the Earth and *The Ecologist* magazine in the late 1960s forced CPRE to respond with a campaign that made the connection between landscape protection and nature conservation. Hedgerows are the most widespread semi-natural habitat in England, relied on by more than 80 per cent of our farmland birds and many rare or threatened species including bats, bees, hazel dormice and, of course, the humble hedgehog. After a 1969 study by Dr Max Hooper estimated hedgerow loss at 10,000 miles a year (as opposed to the Government's official estimate of just 500 miles), CPRE called for 'hedgerow preservation orders' and special protection for the most species-rich hedges. It took over twenty-five years of campaigning before laws preventing the destruction of important hedgerows were introduced in 1997.

Campaigning in the European Parliament throughout the 1980s and 1990s, CPRE helped win payments to encourage farmers to swap destruction for conservation, leading successful campaigns to protect sensitive habitats like Norfolk's Halvergate Marshes from intensive farming in the process. Their work with the Wildlife and Countryside Link continues to ensure that the debate on 'landscape-scale' conservation makes the link between biodiversity and beauty, something that will be crucial for building support for the idea of the countryside as one great nature reserve.

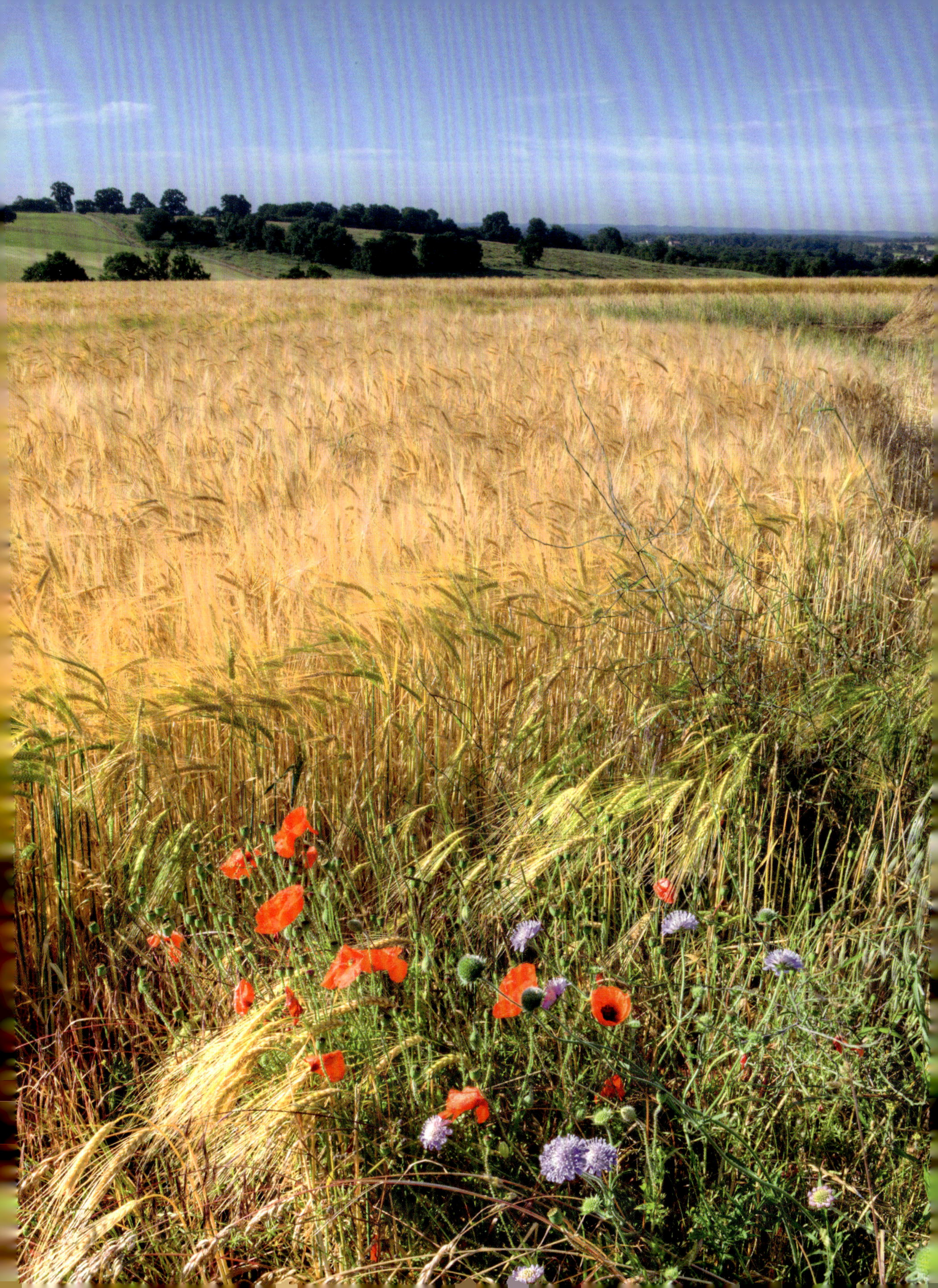

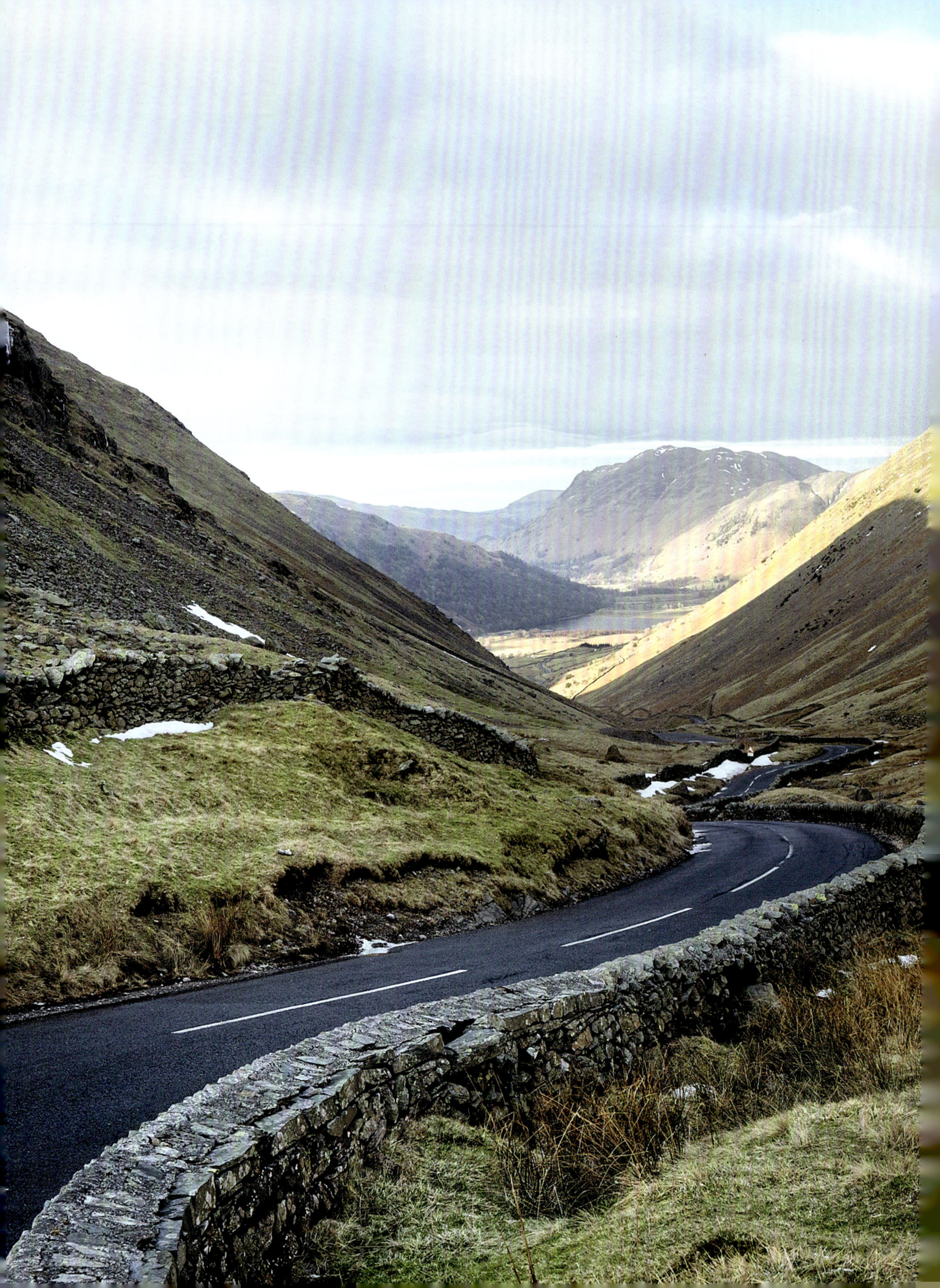

Idea 20

Cutting the Clutter

introduced by Jo Brand

For most of us, a drive in the country is an excuse to escape the bombardment of advertising that we're subject to in the modern city. Let's face it, feasting our eyes on views of the countryside is about the only pleasurable thing about modern driving. Sure, there might be the occasional placard urging us to vote for a certain political party (the price of democracy) or – more helpfully – 'take the next left' for fresh asparagus at the farm shop; but what we have to put up with today is nothing compared to the staggering amount of garish adverts that were strewn along rural roadsides in the first half of the twentieth century. Luckily for us, a determined bunch of principled campaigners made a stand against the scourge that threatened to urbanise the pastoral purity of rural England. Thanks to their perseverance, the countryside remains our last sanctuary from commercialisation.

It was an October 1890 article in the future Poet Laureate Alfred Austin's journal *National Review* that first drew attention to 'The Age of Disfigurement' – the rise of poster and billboard advertising. Having dominated towns for decades, advertising was spreading out to the countryside to capitalise on the increase of visitors made possible by the railways. Richardson Evans' article proposed that legislation was needed to protect 'scenes of remarkable beauty or interest' from the desecration of commercialisation. Two years later, with no sign of progress in sight, the architect Alfred Waterhouse wrote to *The Times* to complain about 'the advertising plague' which meant 'the very loveliest spots in our island are being degraded by this wanton vulgarity'.

Waterhouse offered to set up a society to discourage the trend, and Evans' friends at the Savile Club urged him to make contact. Within days, a meeting was convened at Barnard's Inn in Holborn, the headquarters of the Art Workers Guild, at which the Society for Checking the Abuses of Public Advertising (SCAPA) was formed. The society's journal was named *A Beautiful World* because Mrs Richardson Evans wanted them to be thought of as a Beautiful World Society as much as an Advertisement Regulation Society. William Morris became a founding member of SCAPA, but politely declined to become a public advocate for the cause in a letter of October 1893. Morris felt his potential influence would be limited as someone 'known to be socialist' and that a really respectable person would be a better figurehead, suggesting the Archbishop of Canterbury.

Morris was clear about the size of the challenge ahead, fearing there would not be the 'slightest chance of success' of regulating advertisements without 'a revolutionary act' because 'the advertisements you are speaking of are always on private property'. Despite being disgusted and annoyed by the offensive advertisements 'along the railways', Morris confessed to 'rejoicing at the spectacle of the middle classes so annoyed and so helpless before the results of the idiotic tyranny which they themselves have created'. Undaunted, SCAPA set out on a two-pronged campaign to lobby for statutory regulation while rousing public opinion against the disfigurement of the countryside. Evans and his supporters were concerned not just with the visual nuisance of advertising, but also with the defiling of the sacred purity of England's countryside by rampant commercialisation.

PROFIT IN PRESERVATION

Evans' tireless campaigning eventually succeeded in ensuring that the Advertisements Regulation Act of 1907 was the first in 'the history of English law making to proceed upon the plain principle that the beauty of the landscape is a national asset'. *The Times*' leading article on the same day was wholeheartedly supportive of Evans' arguments on the dangers of putting profit before preservation:

PAGES 186–7
In the 1930s, sections of the Kirkstone Pass were covered in signs advertising the attractions of the Lake District. Thanks to the campaign against roadside clutter, the spiritual heart of the English landscape is no longer spoilt by commercialisation.

SAVE THE COUNTRY SIDE

SAINT GEORGE FOR RURAL ENGLAND

A postcard produced for CPRE's Save the Countryside exhibition of 1928 suggests the organisation felt roadside advertising was the most disfiguring threat to the countryside at that time.

There was a time, and not long ago, when the great mass of educated people seemed to have thought either that beauty was not worth preserving, or else that no exercise of human will could preserve it. Now, luckily, there is a change. We know that it is human will, the will to make money, that destroys the beauty of our countryside; and we begin to see no reason why they should not be opposed by the human will to preserve it... even by the interposition of the State.

But the Act did not actively restrict advertising, merely empowering local authorities to make their own bylaws to that end. Where authorities did wish to remove advertising eyesores, they were stymied by the Act's five-year grace period for existing advertisements. Where bylaws *were* made, those who broke them were not always punished. Despite the dithering of local councillors, the Japanese Ambassador to England, Count Hayashi, was so impressed by the idealism of SCAPA that he returned to Japan and succeeded in introducing a similar law in his own country in 1911.

By 1912, people were beginning to question whether there was any net economic benefit of advertising in the countryside. *The Times* remarked that with the rapidly increasing number of visitors to rural England, 'obtrusive posters or huge field placards are absolutely incompatible with natural beauty' and risked 'the diversion of the tourist stream'. SCAPA was making progress in other areas, convincing the Home Office to rethink their original interpretation of the 1907 Act, which held that the regulations should be confined to special beauty spots. By 1912, Evans had persuaded the Home Office to sanction local bylaws prohibiting disfiguring advertisements within view of any public thoroughfare, including railway lines and rivers, with Hampshire leading the way in using the powers of the Act.

THE RISE OF ROADSIDE ADVERTISING

This limited progress was soon undermined by the onset of war in 1914, and the end of war only brought an acceleration of advertising in the countryside, specifically alongside the roadsides, which, hitherto, had not been considered strategic enough locations for advertisers. As the wartime factory output gradually reverted to peacetime production, the huge improvements in technology and production-line techniques were soon put to use by the motor-car industry. By 1922, there were 183 car producers in the UK, churning out increasingly affordable products for the middle-classes. As these aspiring consumers took to the highways and byways of England, the advertisers responded by plastering roadsides with enticements to buy their products. Lord Newton was moved to remark that year that 'so many persons are obsessed with this passion for advertisement that I am convinced that there are any number who, if they had the chance, would plaster the dome of St Paul's or the bald heads of their middle-aged friends'.

The rise of motoring also created a whole new form of advertising, as the oil companies competed to entice drivers to their garages and service stations. During 1923, Lieutenant Colonel Hobart JP urged the Isle of Wight Chamber

Shell had led the way in removing their roadside signs in 1923; here they are removing signs from a garage in 1928, when the Advertisements Regulation Act was extended to apply to filling stations after CPRE campaigning.

'It is a question of whether we, as a civilised people, really care for beauty.'

of Commerce to lobby Shell-Mex, British Petroleum and Anglo-American Oil about their competing signs on the island. After lengthy negotiations the companies agreed to withdraw signs, while Anglo-American offered a truce on rural advertising throughout the country if their rivals were prepared to match them. The Royal Automobile Club helped SCAPA lobby the oil companies, encouraging British Petroleum to abandon roadside advertising in the Isle of Wight, Cornwall, Kent, and the Lake District. Shell soon announced that 'in sympathy with the movement for the preservation of the natural beauty of the landscape, we are proceeding forthwith to dismantle and remove all our road signs of an exclusively advertising character throughout the countryside'. Anglo-American followed suit, saying: 'In view of the efforts of the SCAPA society to preserve the beauty of the countryside, we have decided to remove all Pratt's Motor Spirit field signs.' Feeling left out, the Dunlop Rubber Company wrote to SCAPA in early 1924 to point out that they had 'decided to renounce approximately 6,000 wayside signboards' as early as spring 1923.

Of course, the motoring advertisements were only a fraction of the total, and a forceful 1924 circular from the Ministry of Transport called on local authorities to address the hoardings that caused motorists to be 'deprived in great measure of the enjoyment which public highways should afford'. A second Advertisements Regulation Act was passed in 1925, prompting the twenty-three-year-old Earl De La Warr, the first hereditary peer to join the Labour Party, to state: 'It is a question of whether we, as a civilised people, really care for beauty; whether we intend to permit individuals actuated purely by commercial motives to vulgarise the natural beauties of our countryside.'

The 1925 Act widened the scope of the 1907 legislation to include any advertisements that might 'disfigure or injuriously affect rural panoramas, the amenities of any village and the amenities of any historic or public building or monument or of any place frequented by the public solely or chiefly on account of its beauty or historic interest'. Yet again, though, it relied on local authorities to create bylaws, and again gave existing advertisements five years' grace before they could be removed. Patrick Abercrombie used *The Preservation of Rural England* of 1926 to argue that although roadside advertisements could be classed as a 'temporary disfigurement', if left unchecked 'they are capable of ruining everything'. Despite the new legislation, and an enlightened public 'fully alive to the enormity of advertisement abuse', Abercrombie noted that 'a very large number of offenders are still flaunting their vile signs'.

THE MENACE OF COMMERCIALISATION

Such offenders prompted a Parliamentary debate in 1928 in which Lord Hunsdon submitted that 'the beauty of the landscape is a great national

> SATURDAY. THE DAILY EXPRESS. MARCH 2, 1929. Moon Rises 0.41 a.m., Sets 9.15 a.m.
>
> HAD JOHN CONSTABLE LIVED TO-DAY.
> The Save-the-Countryside Exhibition is now showing in London.

Strube's *Daily Express* cartoon – 'a long way after Constable' – promotes a 1929 CPRE exhibition highlighting the damage done by advertising.

possession and that it is not right to leave the protection of the landscape to the uncertain policy of local authorities'. Praising the French Government's decision to ban outdoor advertising because 'they realise, as I am afraid we do not, that the beauty and interest of a country is a great national and a great material asset', Hunsdon warned that Shakespeare's 'immortal' descriptions of England were

> . . . rapidly becoming out of date. If you should expect to see some glorious morning 'gilding pale streams with heavenly alchemy', you will find it gilding the most unheavenly pigments of a petrol pump, or if you think to see it 'kissing with golden face the meadows green', you will find it kissing an advertisement of Mr. Carter's Little Liver Pills.

Clough Williams-Ellis' *England and the Octopus*, published by CPRE that year, made 'Advertisements' the first entry in its 'Devil's Dictionary', drawing attention to the 'vulgar Babel' stretching out from London's Bath Road 'almost to Maidenhead'. As well as the promotion of goods, Williams-Ellis noted the countryside was increasingly at risk from 'frantic roadside touting', beginning 100 miles away from a destination and repeated every mile: '100 miles to the Dewdrop Inn: Fried Chicken Lunches . . . 99 miles to the Dewdrop Inn . . .', etc. Williams-Ellis provided a crumb of comfort in revealing that 'even Japan has its advertising afflictions', quoting the Bengali poet and philosopher Sir

Rabindranath Tagore's assessment of the advertising invasion of Japan's pure green fields: '. . . commercialism is a terrible menace to society because it is setting up the ideal of power over that of perfection.'

Back in England, the commercial powers were beginning to realise that with their power came responsibility. Williams-Ellis commended 'the more progressive tyre and petrol firms' who withdrew their rural advertising, but castigated the Raleigh Cycle Company for their garish yellow signs which he felt sure had ruined many a cycle ride. Shamed into action, Raleigh started taking down its signs in 1930. Since 1926, Williams-Ellis and his colleagues at CPRE had given SCAPA a powerful new ally. Indeed, Patrick Abercrombie felt SCAPA was an integral part of the 'little group of unabashed enthusiasts' who created the new organisation, giving them credit for being the first national society that 'sought to look after the country generally'. Their joint campaigning led to the Petroleum (Consolidation) Act of 1928, which applied the principles of the 1925 Act to Petrol-Filling Stations. The legislation meant that local authorities were able to enforce the removal of advertising on and around roadside garages, often located in sensitive tourist destinations and picturesque villages.

THE VILLAGE EYESORE

A CPRE exhibition of 1929 highlighted just how advertising was urbanising England's villages, many of which – despite the success with garage advertising – were afflicted by as many as seventy advertisements for consumer products, with cigarette and tea firms being the worst offenders. The exhibition was

This 1929 hoarding in the Leicestershire countryside prompted Clough Williams-Ellis to say: 'One may remark that advertising sometimes makes strange bedfellows – or is there any connection?'

opened by the leader of the Labour opposition, Ramsay MacDonald, while Mrs Neville Chamberlain hosted an evening reception, indicating the cross-party support for more effective controls of rural advertising. By 1930, the First Commissioner of Works, and soon-to-be leader of the Labour Party, George Lansbury invited CPRE to move its exhibition to Westminster Hall, where even more politicians could be shown how the 1925 Act had failed to prevent the commercialisation of the countryside. Lady Cynthia Mosley MP hoped 'every Member of this House has gone to see that remarkable exhibition in Westminster Hall' where CPRE had 'got together there a really amazing number of instances of the wanton and wholesale destruction' being perpetrated by 'cafés, petrol pumps, garages', for which 'the reckless advertisement is put up, violating the most peaceful and serene landscapes in the most unexpected places'.

In July 1931 Ramsay MacDonald, now Prime Minister, had opened another CPRE exhibition in Chelmsford by arguing 'there was nothing that destroyed the appearance of the countryside more than things which were set up for advertisement purposes ... abominable, vulgar placards, bad in colour, bad in design, and giving bad advice'. He hoped local authorities would 'rigidly and zealously exercise' their regulatory powers, and that the 'great firms responsible' would 'get [advertisements] designed and placed in such a way that they were not an eyesore to every person who loves his country'. The reliance on voluntary action by companies and the creation of local

A country lane in Exmoor just as it should be: completely unadorned by signs or advertising.

bylaws meant progress continued to be slow. The poet and philosopher Sir John Squire told CPRE's 1934 AGM how Sir Herbert Samuel, as Governor of Palestine, had been able to save the landscape of that country ten years earlier by issuing an edict confining advertising to newspapers and railway stations. 'In this country,' he complained, 'the county councils had powers to govern these things, but they did nothing; there were too many vested interests.' By CPRE's 1937 AGM, the Minister of Transport Leslie Hore-Belisha conceded that with 'the observance of the bylaws not yet universal', CPRE and SCAPA had a crucial role to 'stimulate a good taste and right opinion', in the hope of encouraging advertisers to 'respect roadside amenities'.

In 1939, CPRE Lancashire received letters of support from over 100 local authorities after issuing their 'Posters and the Public' report calling for a virtual prohibition of commercial advertising throughout the countryside. As their Secretary Phil Barnes put it, 'it would be difficult to contend that it is in the public interest that young people should grow up in an environment in which they are confronted on all sides with slogans such as "Drink more beer".' To illustrate the report, Barnes drove from Manchester to Lake Windermere, finding 244 adverts on roadsides and bridges; 103 of these were large hoardings carrying several advertisements, while over 500 more adverts were counted on buildings along the route. Though war again put the issue on hold, by February 1945, Lord Mottistone moved to introduce new legislation to ensure 'adequate steps are taken to prevent the disfigurement of the neighbourhood by ugly signs or advertisements'. Mottistone argued 'the returning soldier will rejoice if the countryside is undisfigured and will have his pleasure marred if what he has been dreaming of for so long is ruined by some unsightly advertisement or other sign'.

THE CASE AGAINST OUTDOOR ADVERTISING

If the hopes of returning forces were not enough motivation, Mottistone reminded the Lords that 'we are anxious in the post-war years to encourage foreign visitors to come to this country to see our lovely countryside'. Lord Derwent supported the motion, referring to Phil Barnes' report from Lancashire:

> I suggest that the crux of the matter is to be found in the CPRE report's statement that the advertising industry is a 'wealthy and highly organised one, with representatives throughout the country who before the war were constantly seeking fresh sites, and through its connexion with innumerable other trades has great influence on all local councils'. If this is really so, then it is quite time the local councils ceased to allow themselves to be influenced in favour of encouraging what the report rightly calls this 'quite inexcusable form of disfigurement of which the public generally are sick'.

In 1946, CPRE submitted a lengthy memorandum on 'The Case against Outdoor Advertising' to the post-war Planning Minister, Lewis Silkin, who included provisions to control advertisements in the Town and Country Planning Bill in January 1947: 'It is common ground that outdoor advertisements are capable of doing much harm to amenity. Everyone of normal sensibility knows from his own daily experience, how the more blatant forms of outdoor advertising can spoil natural scenery.' On 19 February 1947, Silkin invited representatives of SCAPA and CPRE to take part in a discussion on the regulations to be framed under the new legislation, thrashing out how Section 31 of the eventual Act would finally provide the effective regulation of advertising.

Reflecting on their campaign, CPRE's 'Posters and the Planning Act' of 1948 declared:

> The coming into force of the new Advertisement Regulations brings to an end the long and sustained struggle by the CPRE to arm the public representatives with effective powers to protect the public highway from this particular disfigurement. During the last hundred years, over 400 Parliamentary bills have been introduced by local authorities seeking to control outdoor advertising. The advertisement sections of the 1947 Town and Country Planning Act finally gave councils powers to prohibit advertisements except on sites approved by them.

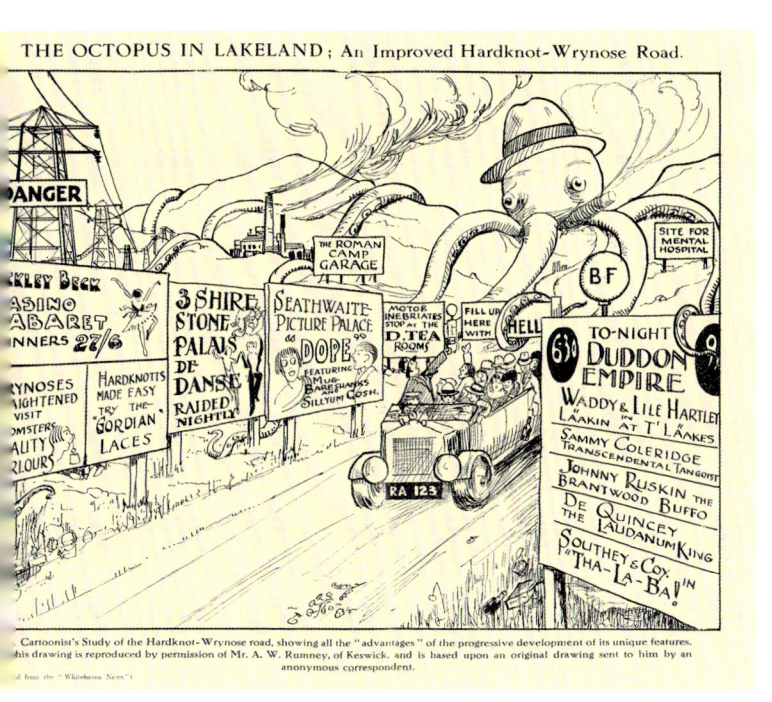

THE OCTOPUS IN LAKELAND; An Improved Hardknot-Wrynose Road.

This satire on the state of advertising in the Lake District (c.1930) reminds us how fortunate we are that we rarely see the countryside of today so badly desecrated by commercialisation. The lamentable situation in parts of the United States and Europe today shows other countries have not been so lucky.

ENJOYING THE VIEW

With SCAPA's objects largely achieved, the society was eventually wound up in 1953, sixty years after its formation, with its advisory work absorbed by CPRE, whose 1954 Annual Review gave the following tribute: 'Scapa can fairly claim much of the credit for the legislation by which outdoor advertising was first bought under control and for the uniform code by which the whole matter is now governed.' The ultimate success of the campaign begun by Richardson Evans was summed up by the then CPRE Chairman, Lord Chorley, in a 1968 debate on the Countryside Bill:

It will be within the recollection of all your Lordships what the countryside used to look like when free advertisement all over the country districts was permitted. Today, as a result of the advertising regulations, our countryside is the most free of advertising and horrid sights of that kind of practically any country in Europe; and certainly this state of affairs contrasts admirably with the situation in the United States. This shows what can be done in the way of removing eyesores.

Gordon Winter's *Country Life* editorial of 1976 pondered that, thanks to CPRE's campaign,

... a whole generation has grown up to whom Ogden Nash's parody, 'I think that I shall never see, A billboard lovely as a tree' is meaningless. Those who are now in their 30s or younger simply do not realise that in their parents' youth it was normal to drive for miles along the main roads out of London and other big cities, without ever seeing an uninterrupted view of the countryside.

More recently, the main form of roadside clutter has been official signs, with CPRE's 'clutter audits' forcing the Transport Secretary himself to concede they were 'a blot on the landscape'; giving CPRE's 2012 Annual Lecture, the Rt Hon Patrick McLoughlin MP admitted that 'ugly and unnecessary signs clutter up the network', sprouting 'like weeds, without any apparent consideration of what's already there'. With new rules issued to address the problem in 2015, the overall situation in England remains far better than in most other countries. However, with the recent trend for adverts on lorry trailers in fields just one example of the ingenuity of those looking to bypass regulations, there remains no room for complacency for the successors of Richardson Evans.

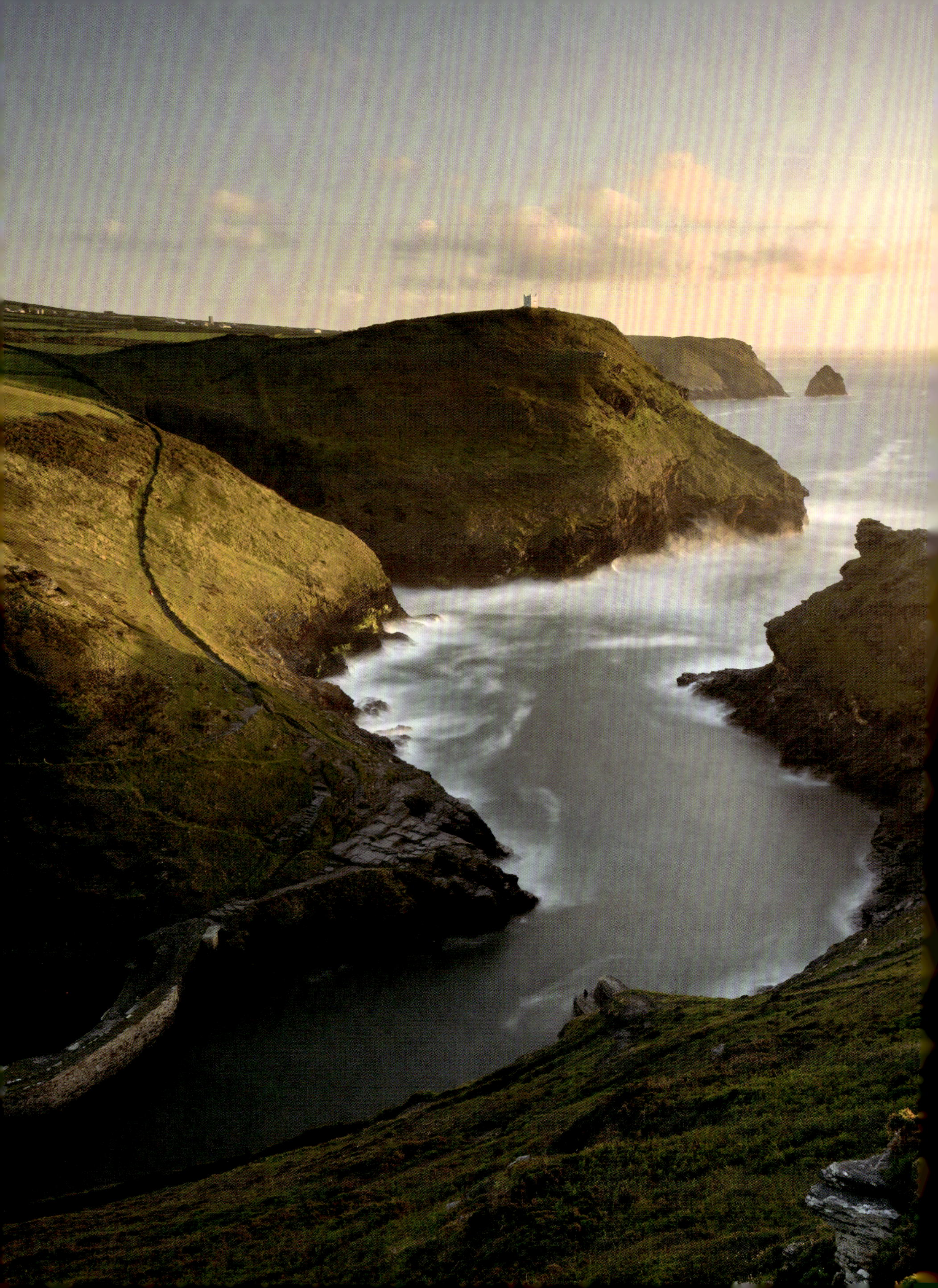

Idea 21

The Coast is the Countryside

introduced by Nicholas Crane

When the BBC started filming *Coast* in 2005, I'd find myself at some point on every shoot bellowing from a headland or beach: 'You're never more than 72 miles from the coast!' It became the programme's slogan. When you think about it, 72 miles is not very far; two or three days for a fit yomper, or a couple of hours by train or car. But for many of us living beyond the mutter of surf, the coast had become a different place; a distant rim beyond the green mosaic of field and wood we knew as 'countryside'. It wasn't always that way. Back when this island was linked to the continent, countryside and coast were connected habitats for foragers and hunters who saw shore and upland as complementary topographies. Harvesting the tideline was just as important as stalking the treeline. During the millennia of development that followed, coast and countryside parted company and it's taken a succession of big ideas to nudge the two back into the same space. Not a moment too soon. Island nations confronted by global challenges need to re-connect edge and centre: field with beach, copse with cliff, brook with bay.

In reviewing Edmund Vale's Batsford guide to *The English Coast* in 1936, *The Spectator* noted that 'most of the guide books to the British Isles fail to communicate the particular pleasure experienced by the coast dweller; they concentrate the attention of the reader exclusively upon the land, overlooking the exciting and beautiful relationship of the land to the sea'. In fact, a year earlier, Batsford's guide to *The Beauty of Britain* had opened with Edmund Vale's chapter 'Coastwise', which nevertheless, went some way to explaining the *Spectator*'s argument: 'I think a great number of people regard our coastline as so much geography. That is a great mistake. It has a subtle beauty about it that has supplied the national mentality with the most powerful elements of its imagination.' Indeed, Peter Ackroyd would later describe the sea as 'perhaps, the true landscape of the English imagination', with the waves 'its own hills and valleys reproducing the soft curves of the English countryside'.

Now, the coast is considered an integral part of the countryside, but for centuries it was treated as a distinctly separate entity to our traditional view of the countryside as rolling fields. Until relatively recently, our coasts have struggled to reach the same levels of appreciation and protection as inland countryside. In his history of British *Coastlines* Patrick Barkham explains that 'the coast was a place for fishing communities, military fortifications and lawless pirates and smugglers. The romantics were in awe of its horror and immensity. The rest of us, even the wealthy, stayed away.' When Cobbett set out on his *Rural Rides* in 1821 he barely touched the coast, only mentioning it to pass comment on shipwrecks and defensive structures (such as the Martello towers, described as circular 'piles of bricks . . . ridiculous things') or to report that Brighton 'is thought by the stock-jobbers to afford a salubrious air'. H.V. Morton, over a century later, charted a similarly inland course in *In Search of England*.

THE RISE OF THE 'SEASCAPE'

Barkham puts his finger on a crucial difference when he says 'the character of our greenest countryside is almost entirely the result of centuries of human activity but the coast is the last repository of wilderness'. As much as Edmund Vale might have protested, the coastline really is 'so much geography', but being formed and shaped by forces outside human control, it soon became an attractive and enigmatic 'seascape' for England's painters. Capitalising on the growing interest in seaside views as a fragile frontier during the Napoleonic wars, the brothers William and George Cooke hired J.M.W. Turner to provide pictures of the south-west coast for their series of engravings, 'Picturesque Views of the Southern Coast of England'. Part of the reason coastal landscapes were increasingly being seen as picturesque was because the landscape tourists were able to reach them far more easily; the boom in turnpike building allowed Turner to complete over 600 watercolour sketches of coastal landscapes around the south-west in the summer of 1811. The resulting views were drip-fed to

PAGES 198–9
The beauty of Boscastle harbour in Cornwall; nowhere is the coast a more integral part of the local countryside.

IDEA 21 • THE COAST IS THE COUNTRYSIDE

J.M.W. Turner's *Hamoaze from St John, Cornwall* was sketched in oils on his second trip to the South West Coast in 1813, and said to have been completed in less than half an hour. Turner's coastal tours would help feed the nation's fascination with seascapes.

the public until 1826, by which time Constable had added coastal scenes to his repertoire and the Norwich School painters like James Stark, John Sell Cotman and John Crome were celebrating the Norfolk coast in oils.

Within two years of Turner's trip to Devon, William Daniell had decided to record the whole of Britain's coastal landscape in what became *A Voyage Round Great Britain* in eight volumes. Previous surveys had mainly been concerned with charting the precise outline of the coast, but Daniell wanted to 'illustrate the grandeur of its natural scenery'. Gradually, the coast became a desirable destination for those seeking out the views captured by Turner and Daniell, rather than the preserve of the upper classes seeking the restorative recreation of sea bathing. The rise of the railways soon meant commercial seaside resorts began to develop in response to the mass demand for coastal holidays. This, in turn, opened up the coast for major development and urbanisation; when Fanny Talbot donated her land above Cardigan Bay to the National Trust in 1896, it was in the hope that it would never be 'vulgarised' by the 'abomination of asphalt paths and cast-iron seats' that had become a common sight around the most attractive sections of Britain's coastline.

PRESERVATION AND RESERVATION

By April 1914, a campaign to create a public path around the Cornish coast was in full swing in response to the march of the holiday bungalows whose gardens stretched all the way to the cliff edge. But despite the efforts of the

National Trust and, from 1926, the Council for the Preservation of Rural England, the situation continued to worsen without a coordinated plan to protect the coast. In 1929, Vaughan Cornish, a renowned geographer and CPRE's leading coastal preservationist, made the case for any future National Parks programme to incorporate England's most beautiful coasts; the creation of these 'seaside parks' would commence with 'the southern extremity of the west coast' of Cornwall, where 'the shore lies open to the majestic swell of the Atlantic . . . one of the finest spectacles of our scenery'. The lack of a continuous, large-scale landscape meant the Cornish coast was not included in CPRE's initial wishlist of twelve National Parks, although Patrick Abercrombie conceded 'there are certain coastal strips, which perhaps have a wider than regional appeal'.

CPRE Cornwall's magnificent 1930 survey of the county drew attention to the need for local authorities to co-operate in upholding and restoring coastal rights of way, suggesting that local Boy Scout groups could 'reconnoitre and patrol' the paths. Most importantly, the survey proposed developing Coastal Reservations to protect the local flora and fauna, drawing attention to the claims of the 'Cornish Riviera' for National Park status with its 'infinite variety of unspoilt cliff scenery . . . devoid of industries'. CPRE's Devon branch produced their own survey in 1932, which warned that Exmouth would soon become a 'seaside suburb of Exeter joined by a strip of roadside building', suggesting that the county's unspoiled coastline should be preserved as 'Areas of Special Landscape Value'.

CPRE's national report on 'The English Coast: Its Development and Preservation' (1936) stressed that a coastline previously seen as a place for fishing, shipping and quarrying was now 'undergoing a rapid transformation'. Author Wesley Dougill reported that while the public had finally realised that 'coastal scenery compares favourably with that of any other part of the country', the new national institution of the seaside holiday threatened to see its greatest asset wrecked by 'wrongful exploitation'. Dougill recommended 'the reservation all round the coast of a belt of open space, free from building', based on the principle that 'it should be available in perpetuity for the benefit of the public'. Vaughan Cornish had already argued that 'the coastline is the special scenic heritage of an island people', and in a 1936 lecture, reasserted that the coast was England's most important landscape:

Effective promotion of the seaside as a leisure destination – including by posters like this 1923 effort from Fred Taylor – meant that coasts were facing extensive development as tourist resorts by the mid-1930s.

We have relatively little mountain or forest wild scenery compared to Europe, but excel in the extent of our coast. In spring-time the Cornish cliff lands are carpeted with wild flowers whose massed brilliance of colour rivals the flowery pastures of the Alps.

SAVING COASTS FOR THE NATION

One person taking direct action to save the Cornish coast was Peggy Pollard, the honorary secretary of CPRE Cornwall and grand-niece of the former Liberal Prime Minister William Gladstone. Pollard, under the pseudonym Bill Stickers, had been part of 'Ferguson's Gang' – masked preservationists who made anonymous donations to the National Trust in the mid-1930s and considered themselves disciples of Clough Williams-Ellis. By 1937, Pollard had gone public with her generosity, and *The Times* reported her gift to the nation of 'the beautiful cliff lands extending for three quarters of a mile from Mullion Cliffs – a region where the Trust has hitherto had no foothold'.

Within a month, the Commons, Open Spaces, and Footpaths Preservation Society was reporting that 'building is proceeding unchecked on many parts of the seaboard, open spaces are being enclosed, cliff paths obstructed, bathing coves barred to the public, and scenery spoilt'. The Society joined forces with CPRE and the National Trust to campaign for stronger protection while, in 1938, Vaughan Cornish published his plan to open up his own *Farm upon the Cliff* in Sidmouth to the public. Cornish called on other coastal landowners and local authorities to unite to 'save our cliff-lands from villadom, preserved in all their beauty for the enjoyment and spiritual benefit of the nation'.

During the Second World War, the English coast became even more accepted as a part of the countryside, thanks to the imagery of the white cliffs of Dover and talk of fighting on the beaches. In 1945, John Dower's recommendations for the Government on National Parks in England and Wales included 'selected parts' of the Cornish Coast in his 'Division A' list. The subsequent 1947 Hobhouse Report of the National Parks Committee relegated the Cornish Coast from its twelve recommendations, citing administrative difficulties while admitting that the 'scenic quality and recreational value' of the Cornish coastline met National Park standards. The committee hoped that 'an alternative method of conservation will be made available under the Conservation Area scheme we propose', and this did eventually lead to the Cornish coast being designated as an Area of Outstanding Natural Beauty (AONB) in 1959 (followed by the North and South Devon Coasts by 1960) with the same protection as a National Park.

The more attractive parts of the coast gradually became designated as AONBs or were included as part of larger National Parks like Exmoor and the North York Moors. However, the 'ordinary' but intrinsically beautiful coastline remained vulnerable as planning legislation failed to recognise its unique attributes and vulnerability. In 1961, CPRE campaigning led to

'The coastline is the special scenic heritage of an island people.'

the Government request that local authorities should survey remaining unspoilt countryside, paving the way for the National Trust's Enterprise Neptune initiative to save the 900 miles of coastline found to be 'worth preserving'. The scheme was a great success, securing over 700 miles of coastline to date, but in 2011 the National Trust joined CPRE in producing a new 'Manifesto for Coasts and Seascapes'.

SHAPING OUR SHORELINES

The new manifesto sought to ensure planning protection for landscape extended to seascapes in an integrated way that recognised the importance of the coast for character of the countryside. With coasts increasingly vulnerable to rising sea levels, the manifesto urged that marine planning should take account of the desirability of protecting good quality farmland, and allow local communities to have a say on decisions to allow 'managed retreat'. In some places, coastlines will inevitably have to retreat and allow the sea to create different – but equally valuable – landscapes and habitats; but with 40 per cent of the world's farmland vulnerable to any rise in sea levels, we would have to think very carefully about losing the fertile countryside of the Lincolnshire and Cambridgeshire Fens which – let us not forget – are internationally unique landscapes, alongside the Somerset Levels and Romney Marsh.

Thanks to CPRE's pioneering coastal campaigners, our seasides are now finally considered part of the countryside, not viewed as an afterthought. When CPRE's President Sir Andrew Motion interviewed England's five main party leaders ahead of the 2015 General Election, four of them chose the coast as their favourite part of the countryside. Prime Minister David Cameron called the coastal path between Polzeath and Port Isaac 'one of the wonders of the world', while Labour leader Ed Miliband said 'When I think of the countryside our wonderful coastline comes to mind first.' With this kind of political support, it is no wonder that coasts are just beginning to attain the same levels of protection and access as inland countryside; a National Trust survey also found that two thirds of the British public say that visiting the coast is important to their quality of life. The work of the Trust, CPRE and others in promoting and protecting the beauty of the coast means that Patrick Barkham is right to conclude that 'our impulse to conserve the coast has shaped our shoreline more profoundly than any destructive endeavour over the last half century'.

OPPOSITE
The stunning beauty of the Cornish coast makes it one of England's most popular 'staycations', and explains why it was the focal point for the early coastal preservationist of the 1930s.

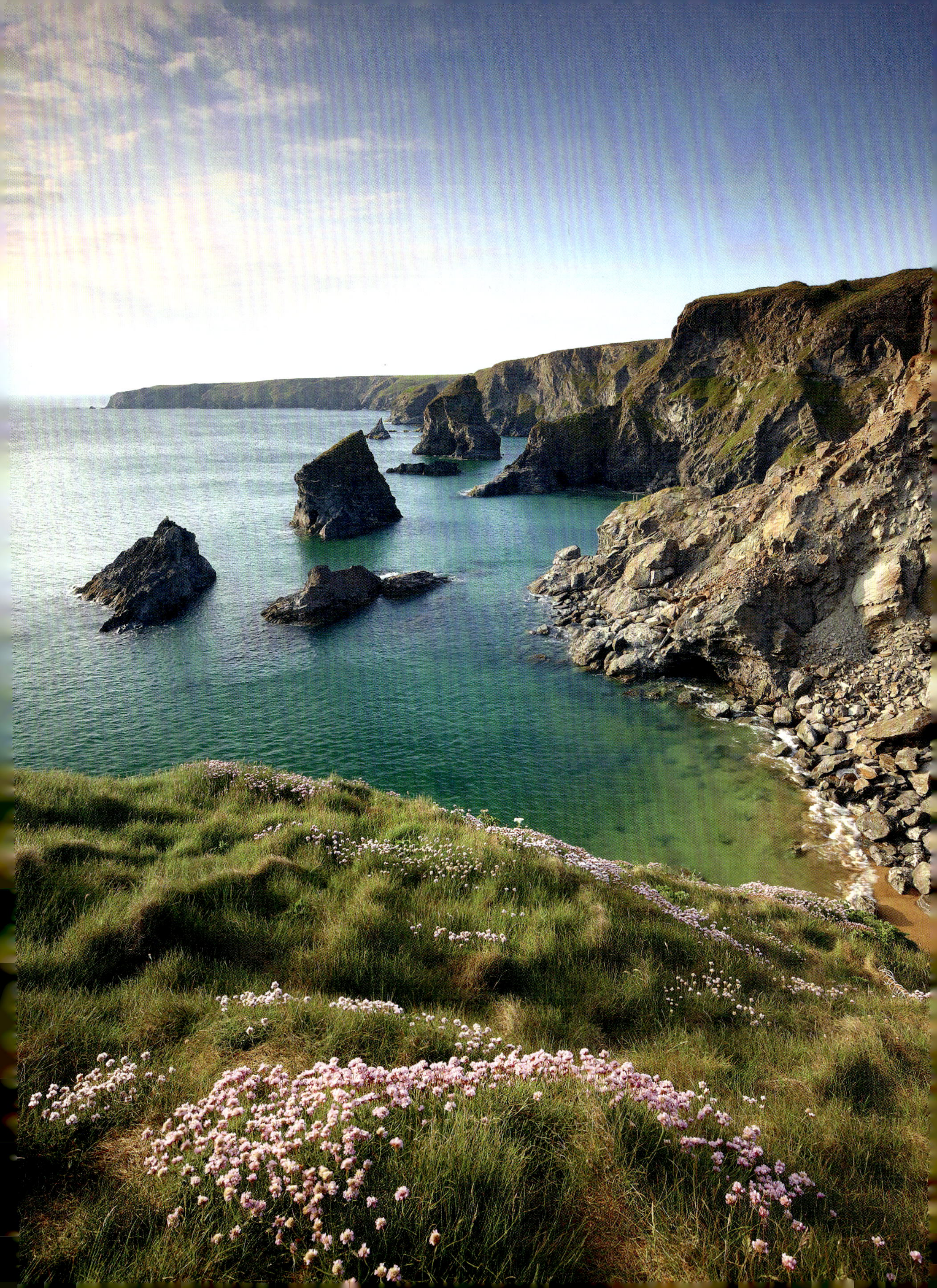

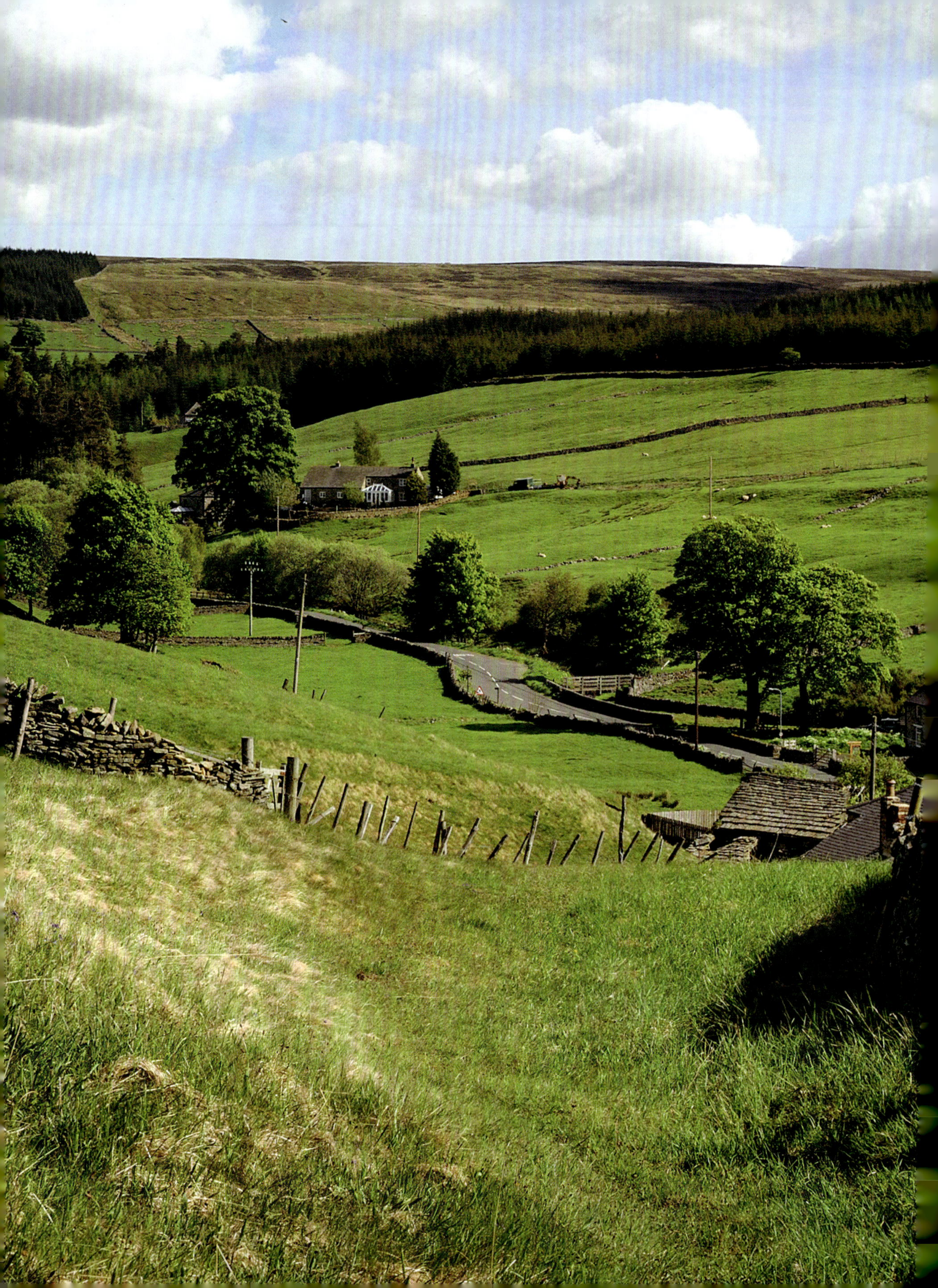

Idea 22

The Country Code

introduced by Caroline Quentin

The Country Code has become part of the culture of England's great outdoors, encouraging townies to enjoy the best of our countryside while treading lightly on the landscape. Over the years, it has given millions of us a better appreciation of rural life and helped create the tourism boom that has allowed our National Parks to thrive. And it's amazing to think that something created for a very specific moment in time – the opening up of the National Parks to the masses in the early fifties – has remained so relevant, opening generations of fresh eyes to the mysteries and customs of the countryside. Above all, it has helped us relearn, as a nation, that our most beautiful landscapes are also places of work for thousands of people: the heroes of the countryside who produce our food and look after our wildlife, despite having to put up with tourists traipsing through their 'offices'!

THE Country Code so many of us grew up with was launched by the National Parks Commission in 1951 and quickly became part of the national consciousness. It was drummed into children in the classroom and on nature walks, while the Country Code badge was much sought after by a generation of Scouts and Guides. The rest of the population absorbed the Code's advice on the responsible enjoyment of the countryside through a mass propaganda campaign; a series of public information films captured the public's imagination, and the iconic poster designs of Norman Thelwell became a mainstay of trains, holiday camps and even factory canteens. The Code marked the State's first acknowledgement of the need for holidaymakers and day-trippers to respect the rural way of life, gently implying that urban standards of behaviour were not always acceptable in the countryside. Harry Batsford's *How to See the Country* of 1945, for instance, retold how his friend had witnessed a 'large party, presumably from the East End' emerge from motor-coaches with a gramophone and proceed to start 'fox-trotting on the turf'.

Patrick Abercrombie had warned about the impact of urban behaviour on rural tranquillity as early as 1933, complaining of 'the honk of the motor-car' and 'the sound of the gramophone' whose 'dissonance is seriously felt and of singular pervasiveness'. In 1930, Vaughan Cornish wrote of the need to educate people to realise that they would 'derive more pleasure from their visits' if only they pursued the 'enjoyment of natural beauty' as much as 'mere jollification'. G.M. Trevelyan's foreword to the Code stressed the importance of the new document in relieving the tension between the countryside's role as the nation's food factory, and in its other main purpose: 'to enable city dwellers upon their holidays to enjoy the refreshment, physical and spiritual, of natural sights and natural sounds.' Trevelyan hoped the Code's 'simple rules of conduct' would stop 'the farmer regarding the holiday maker from town as his enemy' and prevent selfish vandalism destroying 'the pleasure of the next visitors'.

PEACEFUL PERSUASION

The Country Code symbolised the post-war spirit of co-operation and the idea that the countryside we had all fought to protect could now be enjoyed by all. But its origins can be traced back a further twenty years. By the early 1930s, the rise in the popularity of rambling and other rural pursuits among the urban working classes were causing real conflict in the countryside; farmers and landowners were incensed by damaged crops and escaping livestock, while the dropping of litter and mass picking of wild flowers betrayed a lack of respect for their work as stewards of the countryside. In fact, it was the ramblers themselves who first took action, realising that a minority of

PAGES 206–7
A family farm in Allendale, Northumberland, just waiting to be explored by walkers armed with the Countryside Code.

ABOVE
The first edition of the Country Code built on the format of CPRE's earlier Code of Courtesy for the Countryside and urged visitors to 'respect the life of the countryside'.

OPPOSITE
The Forestry Commission took a more hard-hitting approach to encouraging good behaviour, with this 1960s poster warning of 'a horrible death to wildlife'.

culprits would tarnish the reputation of those who, as C.E.M. Joad observed in 1934, had a 'passion for the closing of gates' and a tendency to 'hunt litter like sleuths'. It was in the Peak District around the time of the Kinder Scout Mass Trespass of 1932 that both the Ramblers' Federation and CPRE began to use peaceful persuasion to encourage good behaviour. CPRE Derbyshire's Moor Wardens scheme was started in 1931, with volunteers patrolling Dovedale during the Whitsuntide holiday to put a stop to any littering and any attempts to damage trees, while keeping watch over cultivated land and livestock. The Ramblers began to offer a 'Warden-Guide' service soon after, to help prevent visitors getting lost or injured and to educate them on the need to avoid conduct likely to besmirch the entire rambling movement.

These schemes were just two of many local initiatives that sprang up across England but in 1932 CPRE's Phil Barnes had the idea of creating a nationwide network of wardens with duties defined by a 'code of behaviour'. A renowned climber and expert on countryside access, Barnes knew from personal experience that some form of expert guidance was needed to avoid the potential for confrontation. Supporting his proposals, Sir Lawrence Chubb of the Commons, Footpaths and Open Spaces Society hoped it would encourage landowners to 'remove a deep-rooted feeling of discontent' while encouraging 'townsfolk' to 'learn to know rural England'. CPRE's General Secretary Herbert Griffin worked with the Youth Hostels Association, the Scouts and Guides Associations and rambling clubs, emerging with a national plan for Countryside Wardens to encourage respectful behaviour using a 'Code of Courtesy for the Countryside'. Drawn up by Griffin's assistant Donald Murray, the new code carried the strapline, 'Don't spoil England: Help preserve it' and its main rules were to avoid littering, making noise and disturbing crops or animals. The spirit of the code could be summed up by its instruction that 'When you go leave the place as you found it.'

PREPARING THE PEOPLE FOR THE PARKS

By July 1935, 276 Countryside Warden volunteers, armed with official-looking badges and armbands, were on patrol across the country's beauty spots, enlisted to 'persuade, educate and help the public to keep the countryside unspoilt'. By the outbreak of war, over a thousand were helping visitors to enjoy the countryside safely and courteously. The

Ramblers' own 'code of countryside conduct' had also been widely adopted by its clubs, and was being taught in Cumberland schools by 1939. After the war, the creation of a Government commission to administer National Parks in the areas most commonly patrolled by these voluntary initiatives shifted responsibility to the State. Fearing the new access rights of the 1949 Act would lead to damage to the countryside 'simply through ignorance or negligence', Lord Carrington of the Country Landowners' Association called for an official 'country code of conduct' to gain the 'co-operation and goodwill' of townsfolk. As the Commission's Chair Sir Patrick Duff, put it: 'the process of preparing the parks for the people requires the complementary process of preparing the people for the parks'.

Griffin acknowledged the need for the State to assume responsibility, but noted that 'in all that has been written about the proposed Countryside Code and Wardens scheme it seems to have been forgotten that as long ago as 1935 CPRE launched such a scheme itself'. He paid tribute to those volunteers who had prepared the ground for an official Country Code by selflessly spending their 'weekends and holidays patrolling areas which had been found to need protection'.

The official Country Code was published on 11 May 1951. *The Times* felt that its ten simple maxims had 'a familiar ring and yet are frequently disregarded with widespread disadvantage to country dwellers and subsequent visitors from the towns'. The ten maxims were: 'Guard against all risk of fire; fasten all gates; keep dogs under proper control; keep to the paths across farm land; avoid damaging fences, hedges, and walls; leave no litter;

ABOVE
The Derbyshire Moors were the scene of the first two experiments in influencing good behaviour in the countryside, with CPRE Derbyshire's 'Moor Wardens' scheme followed by the Ramblers' 'Warden-Guide' service.

OPPOSITE
A classic Thelwell poster to promote the launch of the Country Code in 1951.

IDEA 22 • THE COUNTRY CODE

A CODE FOR THE COUNTRY 12

Whenever countrywards we roam
We visit people in their home—
To earn a welcome far and wide
Respect the life of the countryside.

MANY OF THE FARMER'S TOOLS AND MOST OF HIS BUSINESS SUPPLIES ARE
KEPT WHERE WE CAN REACH THEM. BUT THEY ARE AS PERSONAL TO HIM
AS THINGS IN OUR OWN GARDEN OR WORKSHOP.

'A great deal of the future value of life in this island will depend on the degree to which this Country Code is observed.'

safeguard water supplies; protect wild life, plants, and trees; go carefully on country roads; respect the life of the countryside.' The Country Landowners' Association immediately attached 'the greatest importance to putting the ideas of the Country Code into the head of every schoolchild in the land'. The National Farmers Union also wanted to instil the Code in 'the rising generation', praising it for 'setting out the simple terms on which the town can come to the country and be sure of a welcome'. By 1967, 120,000 copies of the Country Code had been sold, with thousands more posters distributed to 28,000 schools and displayed by London Transport and British Railways.

A hugely popular animated film relaunched the Country Code in 1971, but its portrayal of stereotypically ignorant 'townies' abusing the countryside and baiting angry farmers may have been counter-productive for a time. In 1973, the Ramblers' secretary Chris Hall wrote to *The Times* to complain that the 'well publicised' Code 'lays down how the visitor to the countryside should behave', while farmers were habitually blocking or ploughing up paths. With the updated Code failing to demand good behaviour from landowners, Hall asked: 'Is it not time we had a farmers' code on how those who work the land should behave to users of public footpaths and bridleways?' By 1984, the situation had deteriorated to the point that the Ramblers' Association told its members not to feel obliged to respect the Code because of the widespread violations by farmers themselves.

AN ENDURING APPEAL

Despite this loss of credibility among the walking community, the Country Code had succeeded in teaching a generation about rural life; its simple, engaging messages helping to protect England's countryside in the most practical ways. During Parliamentary debates on the Countryside and Rights of Way Act in 2000, Baroness Miller paid tribute to the impact of the Code, which she said had persuaded landowners of the value of embedding its principles 'in the culture of all of those who use the countryside'.

To reflect the Right to Roam introduced by the new legislation, a new version was launched as the Countryside Code in 2004. With farmers and landowners complaining that the old Code had not been effectively publicised for twenty years, a major promotional drive was instigated, using animal characters created by Oscar-winning animator Nick Park. Countryside Agency Chair Pam Warhurst conceded that 'we have all grown up with the Code but we all have different ideas about what it contains', but predicted the new Code's themes of 'respect, protect and enjoy' would encourage a new generation to enjoy the countryside responsibly.

Finally addressing the concerns of ramblers, the new Code introduced advice for landowners, encouraging them to know their responsibilities and to make it safe and easy for visitors to access their land responsibly.

The enduring relevance and appeal of this simple but incredibly powerful idea was proven once again in 2015, when Cumbrian school children received special, green Blue Peter Environment Badges for their project to redesign the Code. Their head teacher explained that the pupils were enthused by thinking about 'what's important when you're outdoors' and the idea of creating something that would help other children 'understand the importance of respecting our countryside'. On launching the original Code in 1951, G.M. Trevelyan predicted that 'a great deal of the future value of life in this island will depend on the degree to which this Country Code is observed'; he would surely be proud that the children of 2016 still instinctively understand this.

The Borrowdale Fells of the Lake District, where thousands of visitors have benefitted from the grounding in the ways of the country given to them by the Code.

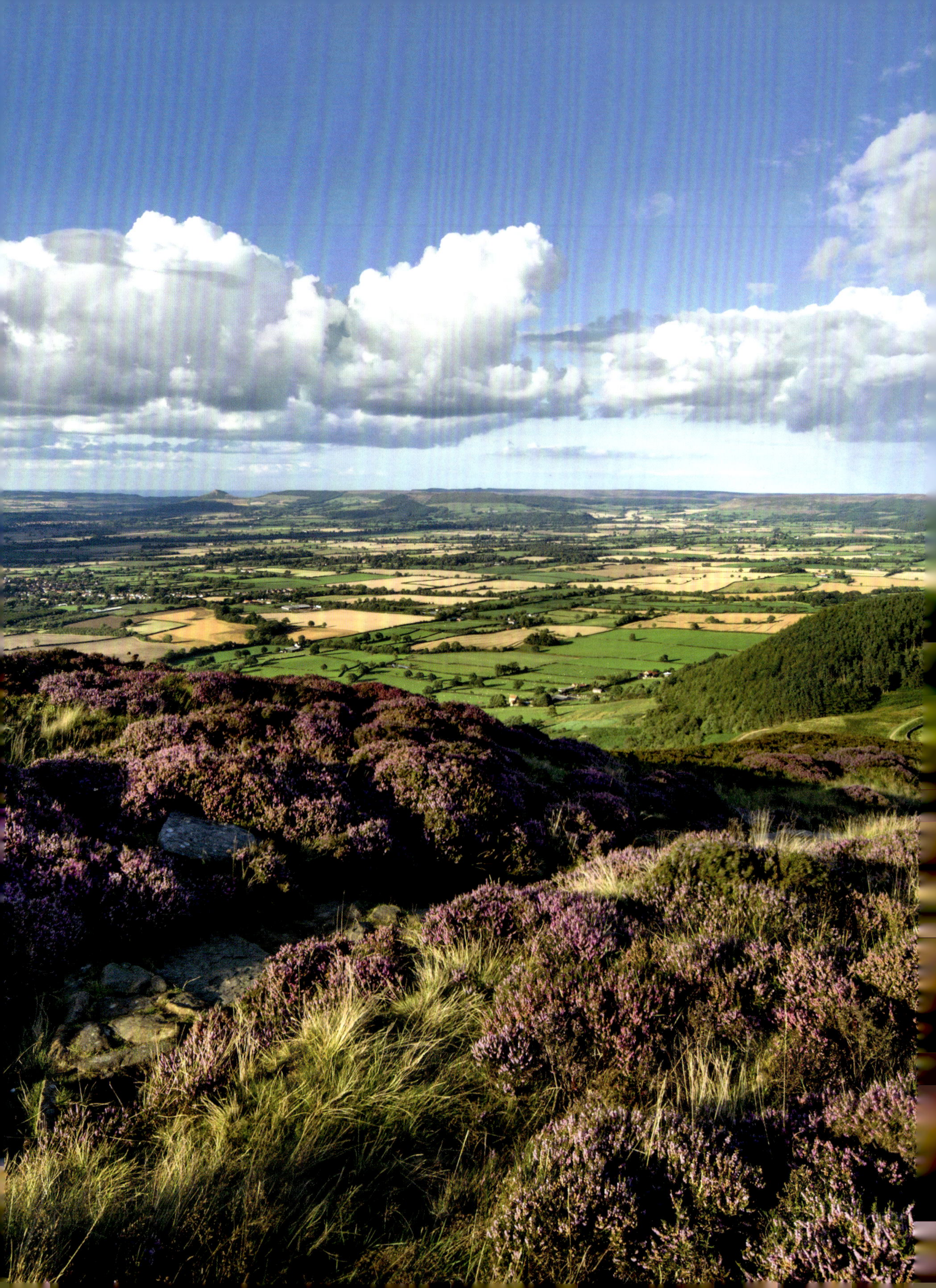

Postscript: Unfinished Business

OPPOSITE
Despite all the successes of our countryside campaigners, even National Park landscapes like Busby Moor in North Yorkshire cannot be taken for granted.

Countryside campaigns are never truly 'won'; they are an ongoing struggle. Unfortunately, we will never be able to regard the achievements included in this book as sacrosanct. That is the nature of campaigning, and all the great pioneers from Wordsworth to Abercrombie knew this. Anyone who cares about the countryside must always remain vigilant; if our guard is ever allowed to drop we can be sure that someone will be making a very persuasive case for building on farmland, or making National Parks about business instead of beauty.

The battle of ideas allows successive generations – people *and* politicians – to re-learn the importance of protecting our countryside, and apply old principles within new contexts. The best ideas are honed through the kind of vigorous debates documented in this book, and their principles will only survive if they work in the modern world. We are confident that the ideas we have chosen are survivors.

In the battle for the countryside, there is a recurring obstacle: the entrenched belief that what is good for the environment must be bad for the economy. The battle between *laissez faire* economics and environmental protection was fought throughout the twentieth century, and many of the great ideas that feature in this book do so because they showed it was possible to balance between the needs of the environment, society and the economy. But this balance is a precarious one; so far in the twenty-first century we have once again seen the economy become the primary factor in any decision involving the future of countryside. Those seeking the honour of high office might do well to remember John Maynard Keynes' argument from 1937's *Britain and the Beast*: governments that use economic grounds to justify the destruction of countryside valued for 'health, recreation, amenity or natural beauty' represent 'a perverted theory of the State'.

While the countryside has historically been defended for its beauty and character, the likes of Keynes always understood its value to our quality of life. David Lloyd George used his Foreword to *Britain and the Beast* to challenge CPRE and the National Trust to make the countryside's 'store of potential health' accessible to all, recognising it as 'a task of supreme importance for our times'. The early preservationist groups often described themselves as 'amenity societies', with Octavia Hill's 1877 essay on *Our Common Land* describing the importance for the 'cooped up' working classes of being able to 'expand into free, uncrowded spaces'.

Since the 1990s, CPRE has led a shift back towards the idea of the countryside as much more than a beautiful heritage, but as a vital source of well-being.

Their research has shown that experiencing rural tranquillity really enhances people's quality of life, but that opportunities to experience it are becoming rarer. In the early 1960s, only a quarter of England was disturbed by the sights and sounds of industrialisation; that figure is now over 50 per cent. And with recent research showing that parts of our cities are 10 times noisier than a decade ago, this is something we need to think long and hard about. Recent polls have shown that seeking out tranquillity is the reason for almost half of all visits to the English countryside, while medical studies are increasingly revealing its importance for improving our mental and physical health. CPRE's tranquillity mapping has shown that natural peace and quiet has been pushed back into small pockets of England, and the growing public concern over this loss has encouraged local and national government to recognise tranquillity as a factor to be considered in planning decisions. However, with the High Speed 2 railway, plans for a new runway for one of the south-east's airports, and major investment in road building and widening, tranquillity is likely to become the main front in the battle to save our countryside.

One aspect of tranquillity where real progress has been made is in the defence of the countryside's star-filled night skies. Back in 2003, CPRE's 'Night Blight' campaign made light pollution a national news story, using pioneering maps based on NASA data to show how it was rapidly spreading out from urban areas. The campaign led to light pollution becoming a statutory nuisance in 2006 and has helped encourage advances in less polluting, energy-efficient lighting. With dark skies linking up so well with the low-carbon agenda and campaigns like WWF's 'Earth Hour' switch-off, CPRE is working with local councils across the country to introduce the new technology, or trial switch-off or dimming schemes. Added to the fact that councils, businesses and householders have all realised that efficient lighting is more cost-effective, there is no reason why we should not aim for a return to 1950s levels of darkness within the next decade. Our children might even get the chance to see the Milky Way with the naked eye, an experience our grandparents took for granted. We can take the fact that Exmoor and Northumberland National Parks recently won protected International Dark Sky Reserve status as evidence of a growing recognition of the need to protect the dark skies we still have. And the growing army of stargazers inspired by Professor Brian Cox give us further cause for optimism.

We've seen how technology can reduce the impact of development in the case of efficient lighting, and sensitive development will be crucial to the future of the countryside. In 2006, the historian Tristram Hunt made the point that 'sustainable development' was at the heart of the anti-sprawl 'message which CPRE has been honing since the days of Abercrombie'. In 2010, the broadcaster Nicholas Crane argued that their campaigns had always been 'beacons of sustainability – leaving our precious soil to store carbon and grow food; minimising waste and pollution; promoting the virtues of local produce to reduce food miles'. And while the ambition of limiting the environmental impact of development is challenging enough, future campaigners must go even further. Many of the ideas in this book show there is plenty of evidence than human intervention can *enhance* the landscape.

Thankfully, we are also beginning to see a growing aspiration to create new buildings of beauty and lasting quality. Most of the great twentieth-century preservationists encouraged positive change in the countryside; in Patrick Abercrombie's words, 'it should be possible for a just balance to be struck between conservation and development' to 'bring forth something new but beautiful'. Anthony Bertram's 1938 Penguin on *Design* explained how for CPRE, preservation did not mean 'turning the countryside into a museum, but rather preserving the life of it' by ensuring that new buildings incorporated the latest innovations in materials and design. He would be pleased to know that CPRE is currently working towards its centenary *Vision for the Countryside* where, in 2026, 'great care is taken to ensure that new buildings enhance local distinctiveness and improve rural areas'.

Ideas like sustainability and quality of life already seem to have been around for an awfully long time without, quite, yet reaching a critical mass. But this book offers nothing but encouragement that if an idea is a good one, it is worth persisting with. And, perhaps in response to globalisation, there is growing enthusiasm for new ideas which champion localism; for instance, Local Food Webs show that protecting the links between small food producers and independent local shops can help rural landscapes *and* economies thrive.

While we hope that new ideas will justify their inclusion in future editions of this book, we must continue to defend the core principles of the ones that have already made it into the pantheon. We have a wonderful rural England as a result of the successes of the preceding chapters; but just as we must not be complacent, we need not feel helpless. After all, David did beat Goliath. The pioneers of the ideas celebrated in this book fundamentally changed the way society saw the world, and transformed our personal relationships with the landscape. If we continue to learn from their great ideas we can be inspired to achieve, in the words of Sir Andrew Motion:

> The change in outlook where beauty and well-being matter; where we aspire to improve the quality, and the equality, of life. Embracing beauty, local character and the enjoyment of green, open spaces can help bring about this change. The fate of England, and the planet, is at risk if we do not. An appreciation of our natural heritage, and a growing environmental consciousness, is not just a joy for ever. It is vital and necessary. It is a part of what makes us who we are.

Index

A

Abbeystead 150
Abercrombie, Lascelles 124
Abercrombie, Patrick 7, 43, 46, 48–9, 52, 67–9, 70, 72–3, 76–8, 89, 91–2, 99, 101, 103, 106, 113, 142, 156, 191, 193, 208, 215–17
Aberdare, Lord 103
Abrahams, Harold 106
access, *see* Right to Roam
Ackroyd, Peter 200
Acland, Francis 39
Addison Committee 100–2
Addison, Joseph 110, 174
Adelaide 85
Adie, Kate 59
advertisements/advertising, *see* Idea 20: 186–97
Advertisements Regulation Act (1907) 188, 190
Advertisements Regulation Act (1925) 191
Alice's Adventures in Wonderland (Lewis Carroll) 143
anti-litter, *see* litter
AONB 92, 106, 151–2, 179, 203
Arcadia (Philip Sidney) 120, 122
Areas of Outstanding Natural Beauty, *see* AONB
Aristotle 84, 92
Aslet, Clive 119
Asquith, Herbert 100
Astor, Nancy 56
Attenborough, David 181
Attlee, Clement 64, 71, 80, 159
Austin, Alfred 188
Ayrton, Acton 29

B

Baines, Chris 149
Baker, Herbert 65
Bakewell, Joan 155
Baldwin, Stanley 47, 49, 55,70, 103, 120
Banstead Downs 88
Barabazon, Reginald (Earl of Meath) 87
Barkham, Patrick 200, 204

Barnes, Phil 166–7, 195–6, 209
Barrington Court 39
Batsford guide to *The English Coast* 200
Batsford, Harry 208
Beach Thomas, William 52
beauty (of landscape), *see* Idea 1: 14–23
Bell, Adrian 127
Bell, Clive 136
Bell, Mary 27
Berkhamsted Common 28–9
Bertram, Anthony 217
Best Kept Village Competition 65
Betjeman, John 114, 127, 138–9, 151–2
Bevan, Nye 64
Birkett, Norman 139
Blake, William 26, 122, 139
Bledisloe, Lord 64, 76, 98
Blomfield, Reginald 114
Bonar Law, Andrew 157
Bonham-Carter, Victor 7, 61, 176
Bossom, Alfred 76
Bowling, George (character in *Coming Up For Air*) 127, 129
Boy Scouts 156–7, 202, 208
Bradbury, Julia 163
Bragg, Melvyn 15
Brand, Jo 187
Brandt, Bill 115
Branson, Richard 159
Brett, Lionel 144
Britain and the Beast (1937) 142, 215
Britten, Benjamin 116–7
Brooke, Rupert 124, 132
Brown, Dr John 20
Brown, Lancelot 'Capability' 21
Browne, Kenneth 143
Browning, Robert 32
Bruegel the Elder, Pieter 17
Bryce, James (Viscount) 38, 96–7
Bryson, Bill 84, 106, 152, 159–60
Buchan, John 71, 89, 166
Buchanan, Colin 141, 143, 145
Buckingham, James Silk 86
Bull, William 87
Bunting, Madeleine 175
Burke, Edmund 110
Burnham Beeches 88

C

cables (electricity), *see* Idea 15: 148–53
Cadbury family 89
Cadbury, George 90
Cadogan, Edward 65
Call and Claims of Natural Beauty (G.M. Trevelyan) 114
Cambridgeshire Fens 38, 41, 183, 204
Cameron, David 92, 204
Campaign to Protect Rural England, *see* CPRE
Cannadine, David 124
Canterbury Tale, A (1944 film) 139
Cardigan Bay 201
Carrington, Lord 210
Carson, Rachel 184
Catlin, George 96
Chamberlain, Joseph 36
Chamberlain, Mrs Neville 135, 194
Chamberlain, Neville 46–7, 52, 73, 89, 135
Charles, Prince of Wales 120
Charter for Ramblers (C.E.M. Joad) 125
Cherry Tree, The (Adrian Bell) 127
Chorley, Lord 197
Chubb, Lawrence 156, 166–7, 209
Churchill, Winston 76–9, 104, 132–3, 135, 138
Clare, John 117, 165, 174
Clark, Kenneth 22, 48, 112, 136
Clausen, George 133
Cley Marshes, 182
clutter, *see* Idea 20: 186–97
Coalition Government (2010–15) 80, 147, 177
coast, *see* Idea 21: 198–205
Cobbett, William 48, 122, 126, 200
Coe, Lord Sebastian 153
Coleridge, Samuel Taylor 165
Collier, Price 124–5
Coming Up For Air (George Orwell) 127
Commons Preservation Society (CPS) 27–33, 36, 44, 47, 124
Commons, Open Spaces and Footpaths Preservation Society, 203, 209
Community Forests 84, 92, 177
Conesford, Lord 145

Constable, John 15, 21–3, 112–13, 115, 153, 192, 201
Constable, W.G. 114–15
Constantius I, Emperor 17
Cooke, William and George 200
co-ordinated campaigning, *see* Idea 4: 42–9
Corduroy (Adrian Bell) 127
Cornish, Vaughan 101–3, 113, 202–3, 208
Cornwall 37, 53, 191, 201–4
Coulsdon Common 88
Council for the Preservation of Rural England, *see* CPRE
Country Landowners' Association 210, 212
Country Life 47, 56, 137, 197
Country Parks 84, 66, 92
Countryside and Rights of Way Act (2000) 212
Countryside Agency 212
Countryside Code/Country Code, *see* Idea 22: 206–13
Countryside Wardens 167, 209–10
Cowper, Henry 28–9
Cowper, William 53, 122, 174
CPRE, *see* Idea 4: 42–9, *also* 11, 40, 51–3, 55, 64–5, 68–73, 75, 78–80, 84, 89–91, 95, 99–100, 102–4, 106–7, 113, 115, 117, 123–4, 131–2, 135–7, 142–7, 149–53, 156–9, 165–7, 169, 174–7, 184, 189–90, 192–7, 201–4, 209–10, 215–17
Crabbe, George 116
Crane, Nicholas 199, 216
Crawford, Lord 46–8, 55, 90, 113, 123, 174–5
Cripps, Stafford 53, 72
Crome, John 201
Crowe, Sylvia 176

D

Dalton, Hugh 105, 139, 158
Daniell, William 201
Dark Skies (light pollution) 216
Darwin, Charles 92
Dawber, Guy 45–8
De La Warr, Earl 76, 191
Death and Life of Great American Cities (Jane Jacobs) 144
Dedham Vale 23, 152
Defoe, Daniel 20, 126
Delius, Frederick 115
Democratic Planning, *see* Idea 8: 74–81, *also* 13

Denman, Lady 139
Derwent, Lord 196
Derwent Hall 167
Derwentdale 137
Derwentwater 38, 40
Development Plans 79, 88, 91
Dickens, Charles 26, 86, 182
Dimbleby, Jonathan 51, 146–7, 177
Dinas Oleu 37
Domicilium (Thomas Hardy) 181
Dougill, Wesley 202
Dower, John 101, 105–6, 175, 203
Dower, Pauline 101, 106
Duff, Patrick 210
Dunbar, Evelyn 138
Dürer, Albrecht 17
Dymock Poets 124

E

Ecologist 184
Economist, The 152
Elgar, Edward 115–16
Elizabeth I 84–5, 90, 92
Elizabeth II 65, 158
Emerson, Ralph Waldo 96
Enclosure 13, 21, 27, 29–30, 165, 168
energy infrastructure, *see* pylons
England and the English from an American Point of View (Price Collier) 124–5
England and the Octopus (Clough Williams-Ellis) 54–5, 69, 192
England Revisited (BBC) 139
England Speaks (1935) 126
England's Glory (1987) 146
English Journey (J.B. Priestley) 126
Enterprise Neptune 204
Epping Forest 27, 29, 33, 88–9
Eugenius 17
Evans, Richardson 38, 44–5, 188–9, 197
Evelyn, John 173–4
Exmoor 101, 194–5, 203, 216

F

Far From the Madding Crowd (Thomas Hardy) 112
Farm upon the Cliff, The (Vaughan Cornish) 203
Farrell, Terry 141
Feeney, Frederick 38
Felpham 122
Festival of Britain 158
Fiennes, Celia 18, 20, 126

First World War 39, 45, 65, 100, 114, 132, 136, 165, 174, 181
Ford, Ford Madox 132
Forest of Dean 98, 175, 177
Forestry Act (1951) 176
Forestry Commission, 174–7
Forestry in the Landscape (Sylvia Crowe) 176
forests, *see* Idea 18: 172–7
Forster, E.M. 142
Freud, Lucian 112
Friends of the Earth 184
Friends of the Lake District 101, 103, 150–2, 175
Frost, Robert 124
Fry, Geoffrey 157
Fry, Roger 136

G

Gainsborough, Thomas 18, 20
garden cities 12, 87–8, 136, 141, 147
Gardiner, Cliver 88
George III 18–19
George V 65, 126, 156
Gibbs, Philip 126
Gibraltar Point 183
Giffard, Constance 70
Giles, Ronald 'Carl' 151
Gilpin, William 20–1, 174
Girl Guides 156–7, 208–9
Gowbarrow Park 40–1
Graham, Robert 96
Grahame, Kenneth 123–4,
Graves, Alderman 89
Great War *see* First World War
Greater London Plan (1944) 91
'green and pleasant land', *see* Idea 12: 118–29, *also* 9, 26, 52, 132, 136, 139, 146
Green Belts, *see* Idea 9: 83–93, *also* 7, 12–13, 46, 51, 147, 217
Greenpeace 184
Greenwood, Arthur, 72
Griffin, Herbert 76, 99–100, 102, 106, 136, 150, 157, 158–9, 209–10
Griggs, Frederick, Landseer 47
Guardian (Manchester) 47–8, 69, 92, 133, 150
Gummer, John (Lord Deben) 146–7
Gurney, Ivor 132

H

Hadow, Grace 61
Hall, Chris 212

Hall, Peter 80
Halvergate Marshes, 184
Hampshire Days (William Henry Hudson) 113
Hampstead Heath 27, 29, 33
Hardy, Thomas 112, 124, 181
Hastings, Max, 48, 75
Hay Wain, The (John Constable) 23, 112
Hayter, Lord 159
Haythornthwaite, Ethel 105, 165
hedgerows 180, 184
Henrion, Henri Kay 160
Hill, Octavia 7, 32, 36–7, 87, 89, 215
Hilton-Young, Edward 55, 71–2
Hinchingbrooke, Viscount, 144
Hinton, Christopher 150
Hobart, Lieutenant Colonel 190
Hobhouse Report 203
Hobhouse, Lord 44
Hogarth, William 122
Holst, Gustav 115
Holy Island (Lindisfarne) 150
Homes, Towns and Countryside 143
Hooper, Max 184
Horder, Lord 103
Hore-Belisha, Leslie 55, 195
Horrobin, Ian 72
Housing and Town and Planning Act (1909) 70
Housman, A.E. 123–4, 132
How to See the Country (Harry Batsford) 208
Howard, Ebenezer 87
Howarth, Henry 96
Hudson, W.H. 113
Hughes, Ted 117
Hunsdon, Lord 191–2
Hunt, Tristram 70, 80
Hunter, Robert 30, 32, 36, 39

I
Imaginary Plan for London of 1907 (Paul Waterhouse) 88
In Search of England (H.V. Morton) 126, 200
inclosure, *see* Enclosure

J
Jacobs, Jane 144–5
Jenkins, Simon 35, 80, 120
Joad, C.E.M. 125, 142, 208
Johnston, Harry Hamilton 44–5
Johnston, Thomas 175

Jones, Thomas 26
Joseph, Delissa 44

K
Kapoor, Anish 142
Kauffer, Edward McKnight 175
'Keep Britain Tidy' 158–60
Kelly, Mary 62
Kennet, Lord 169
Kent 19, 37–8, 125, 177
Keynes, John Maynard 142, 215
Kinder Scout 164, 166–8, 209
King's England (Arthur Mee) 127
Kumar, Satish 95
Kupfer-Sachs, Julius 91

L
Lawrence, D.H. 55
Labouchère, Henry 97
Lady Chatterley's Lover (D.H. Lawrence) 55
Lake District 15, 20–1, 26, 32, 38–9, 40, 44, 96–7, 101, 103, 106, 111–12, 117, 136, 151–2, 165, 174–6, 191, 197, 213
Land, The (Vita Sackville-West) 125
landscape, and beauty of, *see* Idea 1: 14–23
Langland, William 115
Lansbury, George 194
Larkin, Philip 117
Lawson, Nigel 176
Lawson, Jack 150
Le Carré, John 25, 146–7
Leland, John 18, 122
Lewis, George 26
Light Railways Bill (1896) 38
Light, William 85
Lincolnshire Fens 204
Lincolnshire, Lord 62
Linnell, John 26
Listowel, Earl of 72
litter, *see* Idea 16: 154–161, *also* 45, 155–60, 208–10
Lloyd George, David 49, 62, 70, 170, 215
Local Food Webs 217
Local Nature Reserves 92
London County Council 87, 89
London Olympics 2012 12, 120, 153
Lorenzetti, Ambrogio 85
Lorrain, Claude 18, 20, 22
Lynn, Vera 139

M
MacColl, Ewan 166
MacDonald, Ramsey 49, 55–6, 70–2, 98–100, 102, 194
Macmillan, Harold 76, 79
Major, John 169
Malvern Hills, 115–16, 150
Mansion House 62, 147
Mary, Princess Royal 158–9
Maud, John 103
McAlister, Gilbert and Elizabeth 143
McEwen, John 137–8
McKenna, Virginia 179
McLoughlin, Patrick 197
Meacher, Michael 170
Mears, Ray 83
Menuhin, Yehudi 115–16
Metropolitan Commons Act (1866) 29–30
Metropolitan Green Belt Act (1938) 13, 89–90
Miliband, Ed 204
Mill, John Stuart 27, 85
Miller, Baroness 212
Milton, John 175
Minsmere Level, 183
Molson, Lord 145, 169
Monbiot, George 153
Morris, William 19, 23, 27, 30–2, 48, 86, 88, 188
Morrison, Herbert 89–90
Morrison, William 104, 143–4
Morton, H.V. 126, 200
Mosley, Cynthia 71, 194
Motion, Andrew 7, 10, 117, 147, 204, 217
Mottistone, Lord 195–6
Mount, Harry 9
Mount Temple, Lord 73
Murray, Donald 209
Musgrave, George and Theresa 33

N
Nash, Ogden 197
Nash, Paul 115
National Colonisation Society 85
National Cycle Routes, 92
National Farmers Union 212
National Fitness Movement 103
National Footpaths Preservation Society 30
National Forest Parks 175
National Forest, The 177
National Land Fund 105

National Parks and Access to the Countryside Bill and Act (1949) 105–6, 129, 169
National Parks, see Idea 10: 96–107, also 39, 129, 136, 139, 149, 152, 163, 169, 175, 202–3, 207–8, 210, 215–16
National Planning Policy Framework (NPPF) 80
National Playing Fields Association (King George V) 65
National Tree Week 177
National Trust, see Idea 3: 35–40, also 30, 44–5, 47–8, 61, 80, 89–90, 96–7, 103, 142, 149–50, 165, 181, 183, 201, 203–4, 215
Natural England 92, 183
Nature Reserves, see Idea 19: 178–85, also 92
Neville, Charles 68–9
Nevinson, C.R.W. 114
Newbould, Frank 135
New Forest National Park 32, 99, 106, 149, 175
New Towns (New Towns Act, 1946) 141, 143–5
Newbould, Frank 135, 137
Newton, Lord 190
Nicolson, Adam 17
North York Moors National Park 5, 96, 106, 203
Norfolk Broads National Park 13, 106
Norfolk Coast AONB 152
Northumberland National Park 216
Norwich School 201
Novello, Ivor 139

O
O'Hear, Anthony 117
Ofgem 149, 152–3
Olmsted, Frederick 87
Orwell, George 55, 125, 127, 129
Osborne, George 147
Our Common Land (Octavia Hill) 215
Our Island Fortress: There's A Land, A Dear Land (1940 film) 137
Our Record Harvest (1942 film) 137
Oxford Book of English Verse 132
Oxford Preservation Trust 89

P
Pagan Papers (Kenneth Grahame) 123
Park, Nick 212
Parry, Hubert 139
Patinir, Joachim 17

Paton, Dorothy 46
Peacehaven 68–9, 73
Peak District National Park 89, 97–8, 106, 136–7, 149, 156, 164–8, 170, 209
Pepler, George 88
Petherick, Maurice 77
Philosophical Enquiry into the Origin of Our Ideas of the Sublime and Beautiful (Edmund Burke) 110
Pick, Frank 133, 136
Piper, John 48–9, 114, 136
planning system, see Idea 7: 66–73, Idea 8: 74–81
Plimsoll, Samuel 165
Pollard, Peggy 203
Porritt, Jonathan 177
Port Sunlight 142
Powell, Francis 29
powerlines (burying or undergrounding), see Idea 15: 148–53, also 10
Priestley, J.B. 126
Prescott, John 146–7
preservation, see Idea 2: 25–33
Preservation of Rural England, The (Patrick Abercrombie) 43, 46, 48, 52, 68, 99, 113, 156, 191
Principles of Political Economy (John Stuart Mill) 85
Pryor, Francis 9, 80
Public Rights of Way 92, 165–6, 169
Punch 45, 47
Puttnam, David 43, 106, 145–6
pylons, see Idea 15: 148–53, also 11

Q
Queen Elizabeth The Queen Mother 159
Queen's Commemoration Open Space Committee 44
Quentin, Caroline 207
Quiller-Couch, Arthur 132

R
Radnor, Lord 144
Rake's Progress (William Hogarth) 122
Ramblers' Association 105, 149, 163, 166, 168–9, 209–10, 212–13
Rawnsley, Eleanor 156
Rawnsley, Canon Hardwicke 32, 36
Read, Herbert 136
Recording Britain (1940) 136–7
Reith, Lord 104
Restriction of Ribbon Development Bill and Act (1936) 56

Reveries of a Solitary Walker (Jean-Jacques Rousseau) 110
Rew, Henry 61–2
Reynolds, Fiona 67
ribbon development, see Idea 5: 50–57, also 68, 144
Ridley, Nicholas 146
Right to Roam, see Idea 17: 162–171, also 7, 212
Rights of Way Act (1932) 166
Rights of Way Act (1990) 169
Robinson, Tony 131
Rogers, Richard 147
Romney Marsh 204
Rothman, Benny 166
Rothschild, Charles 181–4
Rousseau, Jean-Jacques 110–11
Royal Institute of British Architects (RIBA) 45
RSPB 113, 183
Rural Community Councils 61–2
Rural England: The Case for the Defence 104
Rural Planning, see Idea 7: 66–73
Rural Rides (William Cobbett) 48, 122, 200
Ruskin, John 7, 26, 30, 32, 37, 86–7, 113, 122, 173–4

S
Sack of Bath, The (Adam Fergusson) 70
Sackville-West, Vita 125
Salisbury Cathedral 13, 137
Samuel, Herbert 195
Sandby, Paul and Thomas 18–20
Sandys, Duncan 91–2
Savage, William 62
Savile Club 188
Scenery of England (Vaughan Cornish) 103
Schlich, Wilhelm 174
SCAPA (Society for Checking Abuses in Public Advertising), see Idea 20: 186–97, also 44, 47
Scott Report 182
Scott, Leslie 70
Scott, Peter 138
Scott, Lord Justice 104
Scruton, Roger 55, 120
Second World War, see Idea 13: 130–9, also 56, 65, 76, 90, 115, 142, 175, 184, 203
Selborne 33
Selborne League 33

Selborne Society 33
Shakespeare, William 7, 120, 192
Sharp, Thomas 69, 78, 88, 90
Shaw-Lefevre, George (Lord Eversley) 27–9
Sheffield Daily Telegraph 156
Shell Guides 127
Shepheard, Peter 77
Shoard, Marion 78, 80, 165, 170
Shore, Peter 145
Shropshire Lad, A (A.E. Housman) 125, 132
Sidney, Philip 120–1
Silent Spring (Rachel Carson) 184
Silkin, Lewis 78, 106, 144, 169, 196
Silver Ley (Adrian Bell) 127
Simon, Ernest 76
Slaney, Robert 30–1
Society for the Promotion of Nature Reserves (SPNR) 181, 183
Somerset Levels 204
South Downs 101, 106–7, 136–7, 151
Spectator, The 33, 44, 47, 61, 110, 182, 200
Spender, Stephen 150
Sphere, The 87
Squire, J.C. (John) 49, 195
Stamp, Dudley 144, 175
Stephenson, Tom 105, 165, 168
Stevenage, 144
'Stop the Drop' campaign 159–60
Stratford, London 12, 140, 147, 153
Strube, Sidney 192
Suffolk 116–7, 127, 183
Sylva (John Evelyn) 174

T
Tagore, Rabindranath 193
Talbot, Fanny 37, 201
Tansley, Arthur 182–4
Taylor, Fred 202
Tennyson, Alfred, Lord 32–3
Thatcher, Margaret 145, 159
Thelwell, Norman 208, 210–11
This Land Is Our Land (Marion Shoard) 165
Thomas, Edward 124, 132–3
Thomas, Keith 17
Thompson, Piers 98
Tidy Britain Group 159
Tillingham Hall 92
Times, The 39–40, 44–5, 48, 53, 55, 62, 65, 71, 84–5, 89, 97, 115, 127, 135, 151, 156, 167, 174, 188–9, 203, 210, 212
Town and Country Planning (Patrick Abercombie) 103
Town and Country Planning Act (1909) 70
Town and Country Planning Act (1932) 72, 76
Town and Country Planning Act (1947) 12, 78–80, 91, 144, 196
Town and Country Planning system, *see* Idea 7: 66–73, Idea 8 74–81, *also* 51
Town Planning Institute 46, 73
Toys Hill 37–8
Traffic in Towns (Colin Buchanan) 143, 145
Tree Council 177
Trevelyan, Charles 97, 101
Trevelyan, G.M. 20–1, 90, 97, 102–4, 114, 117, 165, 175, 208, 213
Trevelyan, Mary 97, 101
Turner, J.M.W. 15, 22, 200–1

U
undergrounding (powerlines or cables), *see* Idea 15: 148–53
Unwin, Raymond 64, 88
urban regeneration, *see* Idea 14: 140–7, *also* 12

V
Vale, Edmund 200
Vaughan Williams, Ralph 115–16
Vermuyden, Cornelius 13
Victoria, Queen 31, 44, 87
Victoria model town 86
Village Clubs Association (VCA) 61–2
Village Drama Society 62
village halls, *see* Idea 6: 58–65
Virgil 120
Vision for the Countryside, A 217

W
Waine, Peter 137
Wainwright, Alfred 163, 165
Wakefield, Arthur 39
Wakefield, Edward Gibbon 85
Wallhead, Richard 52
Walpole, Horace 19–20
Walton, Izaak 122
Walton Hall, 181
War Memorials 65
Warhurst, Pam 212
Waterhouse, Alfred 188
Waterhouse Paul 88
Waterton, Charles 179, 181
Waugh, Evelyn 55
Wells, H.G. 88, 168–9
West, John Walter 133
Westminster, Duke of 32
Westminster Hall 71, 194
White, Gilbert 33
Wicken Fen 38, 40, 181
Wildlife and Countryside Link 184
Wildlife Trusts 181, 183–4
Wilkinson, Ellen 52
Wilkinson, Joseph 97
Willett, Henry 181
Williams-Ellis, Clough 23, 55, 69, 103, 142, 192–3, 203
Wimbledon Common 27, 29, 33
Wind in the Willows, The (Kenneth Grahame) 123
Winter, Gordon 56, 197
Winterson, Jeanette 173
'wirescape', *see* Idea 15: 148–53
Wombles, The 159, 160
Women's Institutes 47, 62, 64, 156–9
Women's Land Army 139
woodland, *see* Idea 18: 172–7
Woodland Trust 177
Woodwalton Fen 181, 183
Wordsworth, William 7, 15, 22–3, 26, 30, 45, 96–7, 107, 109–17, 165, 168, 174, 215
Wordsworth's 'Sense Sublime', *see* Idea 11: 108–17, *also* 7
Wyld, William 31

Y
Yellowstone National Park 96–7
YMCA 133
Youth Hostel Association 167, 209
Yorkshire Dales National Park 7, 102, 128–9, 170
Younger, George 99

Picture acknowledgements

COVER: Ladybower from Bamford Edge, Peak District National Park, Derbyshire © Tom Mackie Images; BACK COVER: Cleveland Hills, North York Moors National Park © Lizzie Shepherd Photography; TITLE PAGE: Marshwood Vale, Dorset © Guy Edwardes Photography Ltd; 4 Farndale, North York Moors © Joe Cornish Gallery; 6 Gunnerside, Swaledale, North Yorkshire © Lizzie Shepherd Photography; 8 Dartmoor, Devon © David Hughes/Shutterstock.com; 10 © ESA; 11 © David Iliff/Shutterstock.com; 13 © CPRE; 14–15 Colmers Hill, Dorset © Guy Edwardes Photography Ltd; 16 Buttermere, Cumbria © Damian Shields Photography; 18 Royal Collection Trust/© Her Majesty Queen Elizabeth II, 2015; 19 © Peter Watson; Goudhurst, Kent; 21 © Tom Mackie Images; 22 Warren Collection – William Wilkins Warren Fund, Museum of Fine Arts, Boston; 23 Willy Lott's cottage, East Bergholt, Suffolk © Chris Howe; 24–5 The Newlands Valley – saved from a slate railway following one of the earliest campaigns for the preservation of the Lake District in 1883 © Chris McIlreavy; 27 University of Bristol Library, Special Collections; 28 © Helen Dixon Photography; 31 Royal Collection Trust/© Her Majesty Queen Elizabeth II, 2015; 32 © Mark Bauer; 34–5 Cheddar Gorge, Somerset - saved from quarrying operations by the National Trust as one of their earliest acquisitions © Fran Halsall Photography; 37 © National Trust Images/Andrew Butler; 38 Skiddaw and Keswick, Cumbria © Fran Halsall Photography; 40 © NTPL/Rod Edwards; 41 © Joe Cornish Gallery; 42–3 Painswick valley, near Stroud, Gloucestershire © Martin Fowler/Shutterstock.com; 45 *Punch* cartoon, reproduced in Clough Williams-Ellis, *England and the Octopus* (London, 1928); 46 London Transport Museum; published by Underground Electric Railway Company Ltd (1926); 47 © CPRE; 48 Cover of Patrick Abercrombie, *The Preservation of Rural England* (Liverpool, 1926). CPRE Archive, Museum of English Rural Life; 49 above © CPRE (CPRE Archive, Museum of English Rural Life); 49 below © CPRE (CPRE Archive, Museum of English Rural Life); 50–1 Cotswolds © Matthew Dixon/Shutterstock.com; 53 Dawn in West Penwith, Cornwall © Helen Dixon Photography; 54 Cover of Williams-Ellis, *England and the Octopus* (London, 1928) republished by CPRE in 1996 by kind permission of the Portmeirion Estate; 56 © CPRE (CPRE Archive, Museum of English Rural Life); 57 From *The Face of the Land* (1931), Yearbook of the Design and Industries Association 1929–30 (London, 1930); 58–9 © David Crosbie/Shutterstock.com 60 © Chris Howe; 63 above Milkman crossing the River Darent, Kent © CPRE (CPRE Archive, Museum of English Rural Life); 63 below Location unknown © CPRE (CPRE Archive, Museum of English Rural Life); 64 © Tom Mackie Images; 65 © Victoria and Albert Museum, London (given by the Pilgrim Trust); 66–7 Cleveland Hills evening light, North York Moors National Park © Lizzie Shepherd Photography; 69 South Coast Land & Resort Company/Gordon Volk (1922); 71 From *The Face of the Land* (1931), Yearbook of the Design and Industries Association 1929–30 (London, 1930); 72–3 © Chris Howe; 74–5 Bath © Chris Howe; 77 Illustration from Patrick Abercrombie, *Greater London Plan 1944*, (London, 1945); 78 Cover of Thomas Sharp, *Town Planning* (Harmondsworth, 1940); 81 Totnes Castle © Samot/Shutterstock.com; 82–3 Hertfordshire Green Belt © Chris Howe; 85 Ambrogio Lorenzetti, *The Effects of Good Government in the Countryside* (c.1339), Sala della Pace in the Palazzo Pubblico, Siena, Italy; 86 Clent Hills, Worcestershire © David Hughes/Shutterstock.com; 89 London Transport Museum, published by London Transport (1934); 91 Illustration from S.P.B. Mais, *Fifty Years of the L.C.C.* (Cambridge, 1939) © Cambridge University Press; 93 Hertfordshire Green Belt © Chris Howe; 94–5 Cleveland Hills summer, North York Moors National Park © Joe Cornish Gallery; 96 Courtesy of The Wordsworth Trust, Grasmere; 98 © Tom Mackie Images; 99 Bratley View, New Forest, Hampshire © Helen Dixon Photography; 101 Exmoor National Park © Chis Howe; 102 © Stephen Garnett; 107 View from the Devils Dyke, West Sussex © Helen Dixon Photography; 108–9 Scafell Pike, Lake District National Park © Joe Cornish Gallery; 111 © Nick Bodle; 113 © Victoria and Albert Museum, London (bequeathed by Isabel Constable, daughter of the artist); 116 © Chris Howe; 118–19 'The Manger', Uffington, Oxfordshire © Helen Dixon Photography; 121 Win Green © Charlie Waite Photography; 122–3 Wenlock Edge and Brown Clee Hill, Shropshire Hills Area of Outstanding Natural Beauty © Fran Halsall Photography; 124 © CPRE (CPRE Archive, Museum of English Rural Life); 127 © Chris Howe; 128 © Derry Brabbs; 129 © CPRE (CPRE Archive, Museum of English Rural Life); 130–1 The Marshwood Vale from Pilsdon Pen, Dorset © Guy Edwardes Photography Ltd; 133 left © IWM (Art.IWM PST 0320), published by the Parliamentary Recruiting Committee (1915); 133 right London Transport Museum, published by the Underground Electric Railway Company Ltd (1917); 134 London Transport Museum/The Brothers Warbis, published by the Underground Electric Railway Company Ltd (1917); 135 © IWM (Art.IWM PST 14887), published by the Army Bureau of Current Affairs (1942); 137 Kenneth Rowntree, *Grainfoot Farm, Derwentdale* © Victoria and Albert Museum, London (given by the Pilgrim Trust); 139 © Tate, London 2008 (presented by the War Artists Advisory Committee in 1946); 140–1 The Olympic Park in Stratford, London, featuring the Orbit sculpture created by CPRE's Charter supporter Sir Anish Kapoor, Stratford, London © Chris Howe; 143 Illustration from Colin Buchanan, *Traffic in Towns* (London: HMSO, 1963); 146 Salford, Greater Manchester © Chris Howe; 148–9 Pylons, East Sussex © David Iliff/Shutterstock.com; 151 CPRE Archive, Museum of English Rural Life © Carl Giles Estate/Express Newspapers; 153 © Chris Howe; 154–5 Stanage Edge, Peak District National Park © Robert Canis; 157 CPRE Archive, Museum of English Rural Life © The Daily Mail; 158 © CPRE (CPRE Archive, Museum of English Rural Life); 160 © CPRE; 161 © The Estate of FHK Henrion, courtesy of the FHK Henrion Archive, University of Brighton Design Archives (with the permission of the Henrion Estate); 162–3 Causey Pike, Cumbria © Chris McIlreavy; 164 High Cup Gill near Dufton, Cumbria © Chris McIlreavy; 167 CPRE Archive, Museum of English Rural Life © The Times; 168: © Fran Halsall Photography; 169 CPRE Archive, Museum of English Rural Life © The Ramblers' Association; 170 © Darren Ciolli-Leach; 171 © Lizzie Shepherd Photography; 172–3 Micheldever Wood (managed by the Forestry Commission), Hampshire © Fran Halsall photography; 175 London Transport Museum, published by the Underground Electric Railways Company Ltd (1932); 176 Loughrigg near Grasmere, Cumbria © Ross Hoddinott Photography; 177 CPRE Archive, Museum of English Rural Life; 178–9 Elmley National Nature Reserve, Kent © Robert Canis; 180 Clattinger Farm Nature Reserve near Somerford Keynes, Wiltshire© Wolstenholme Images/Alamy Stock Photo; 182 © Robert Canis; 184 © CPRE (CPRE Archive, Museum of English Rural Life); 185 Benington, Hertfordshire © Chris Howe; 186–7 Kirkstone Pass, Lake District National Park © Chris McIlreavy; 189 © CPRE (CPRE Archive, Museum of English Rural Life); 190 © CPRE (CPRE Archive, Museum of English Rural Life); 192 CPRE Archive, Museum of English Rural Life © Strube/Express Newspapers; 193 © CPRE (CPRE Archive, Museum of English Rural Life); 194–5 Parsonage Farm, Exmoor National Park, Somerset © Tom Mackie Images; 197 CPRE Archive, Museum of English Rural Life; 198–9 Boscastle harbour, Cornwall © Ross Hoddinott Photography; 201 Tate, London 2011; 202 London Transport Museum, published by the Underground Electric Railways Company Ltd (1923); 205 Bedruthan Steps, Cornwall © Helen Dixon Photography; 206–7 Allendale, North Pennines, Northumberland © Chris McIlreavy; 208 Front cover of the *Country Code* (London: HMSO, 1951); 209 CPRE Archive, Museum of English Rural Life; 210 © Chris Howe; 211 Norman Thelwell, *Code for the Countryside* poster 12, prepared for the National Parks Commission by the Central Office of Information; CPRE Archive, Museum of English Rural Life; 213 © Tom Mackie Images; 214 © Lizzie Shepherd Photography

First published in 2016 by Frances Lincoln Ltd
74–77 White Lion Street
London N1 9PF

QuartoKnows.com
Visit our blogs at QuartoKnows.com

22 Ideas that Saved the English Countryside
Copyright © Frances Lincoln Ltd 2016
Text copyright © Peter Waine and Oliver Hilliam 2016
Copyright of each of the introductions rests with the author under whose name it appears
Illustrations copyright see page 223

All rights reserved. No part of this publication may be reproduced, stored in a retrieval system or transmitted, in any form, or by any means, electronic, mechanical, photocopying, recording or otherwise, without either prior permission in writing from the publishers or a licence permitting restricted copying. In the United Kingdom such licences are issued by the Copyright Licensing Agency, Barnard's Inn, 86 Fetter Lane, London EC4A 1EN.

ISBN: 978-0-7112-3689-9

Printed and bound in China

9 8 7 6 5 4 3 2 1

Quarto is the authority on a wide range of topics.
Quarto educates, entertains and enriches the lives of our readers – enthusiasts and lovers of hands-on living.
www.QuartoKnows.com

Develop Your Maths

Handling data

Level 1

nec
NATIONAL
EXTENSION
COLLEGE

Handling data Level 1

© 2005 National Extension College Trust Ltd. All rights reserved.

ISBN 1 84308 314 0

Author:	Joanne Hockin, Numeracy and Maths Tutor
Consultant:	Chrissie Minton
Project Manager:	Steve Attmore
Copy Editor:	Jean Twine
Proof-reader:	Andrew Johnston
Cover image:	GOODSHOOT/Alamy
Page design by:	John Matthews
Page layout by:	John and Deborah Matthews
Printed by:	Pear Tree Press Ltd.

Every effort has been made to contact the copyright holders of material reproduced here.

> No part of this publication may be reproduced, stored in a retrieval system, or transmitted in any form or by any means, electronic, mechanical, photocopying, recording or otherwise, without the prior permission of the publisher.
>
> This publication is sold subject to the conditions that it shall not, by way of trade or otherwise, be lent, resold, hired out, or otherwise circulated without the publisher's prior consent in any form of binding or cover other than that in which it is published and without a similar condition being imposed on the subsequent purchaser.

The National Extension College is an educational trust and a registered charity with a distinguished body of Trustees. It is an independent, self-financing organisation. Since it was established in 1963 NEC has pioneered the development of flexible learning for adults. NEC is actively developing innovative materials and systems for open and distance learning opportunities on over 150 courses, from basic skills to degree and professional training.

For further details of NEC resources contact:

National Extension College Trust Ltd
The Michael Young Centre
Purbeck Road
Cambridge CB2 2HN

Tel 01223 400300   Fax 01223 400321
Email: resources@nec.ac.uk   website: http://www.nec.ac.uk

Registered charity 311454

Contents

Introduction

Section 1: Using data and statistical measures
    Topic 1: Understanding data
    Topic 2: Collecting and showing data
    Topic 3: Finding the arithmetical average
    Topic 4: Finding the range

Section 2: Probability
    Topic 1: Introduction to probability
    Topic 2: Measuring probability

Glossary

Information on the National Test

LONDON BOROUGH OF CAMDEN	
4273729	
Bertrams	22.02.08
	£15.00
F	

Introduction

Welcome to the workbook on handling data. This workbook will help you to read and understand information presented in the forms of:

- tables
- diagrams
- charts
- line graphs.

It will also help you to analyse information and understand probability.

In real life, everywhere we go we are surrounded by information. This could be in the form of:

- newspapers
- television programmes
- magazines
- timetables.

All of which can be confusing. By helping you to understand how data is presented, this workbook will help you make more sense of such information.

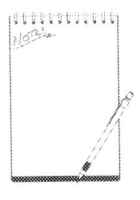

Resources

For your work in this workbook you will need the following:

	Section 1	Section 2
Notepaper and pen/pencil	✔	✔
Graph (or squared) paper	✔	
A ruler	✔	✔
Protractor (for measuring angles)	✔	
A pair of compasses (or something to draw a circle with)	✔	
A holiday brochure or a newspaper	✔	
A map	✔	
A set of dominoes or a coin		✔
A die		✔

The national standards for adult numeracy

In Autumn 2000 the government launched a national strategy to improve the numeracy skills of adults in England. The strategy includes:

- national standards for adult numeracy

- a core curriculum to show what learners need to know in order to reach those standards

- a new system of qualifications in line with the standards

- better learning opportunities to meet the needs of a wide range of learners.

The standards describe adult numeracy as 'the ability… to use mathematics at a level necessary to function at work and in society in general'.

The standards provide a map of the range of skills that you are expected to need.

Numeracy covers the ability to:

- understand and use mathematical information
- calculate and manipulate mathematical information
- interpret results and communicate mathematical information.

Level 1 of the national adult numeracy standards is equivalent to Key Skills Level 1 and NVQ Level 1.

Aims of the workbook

The aims of the workbook are to provide:

- examples of numeracy from everyday life, with relevant methods shown
- lots of learning activities in different practical contexts.

By the time you have completed this workbook you will be able to:

- extract discrete data from tables, diagrams, charts and line graphs
- collect, organise and represent discrete data from tables, diagrams, charts and line graphs
- find the mean and range for a set of data
- show that some events are more likely to occur than others
- express the likelihood of an event using fractions, decimals and percentages with the probability scale of 0 to 1.

This workbook is part of the NEC's 'Develop Your Maths' series. The series has been produced in response to a need for materials for Level 1 numeracy.

The other workbooks in the series are:

- Number
- Measures, Shape and Space.

If you would like to get the Level 1 Certificate in Adult Numercy, you should work through all three workbooks or be confident in the skills covered before taking the Level 1 National Test. You can find out more about this test in the section 'Information on the National Test' at the back of the workbook.

Who the workbook is for

Over 7 million adults in England have difficulty with numeracy. This means they have difficulty doing tasks such as calculating how much change they should get or working out the equivalence between a fraction and a decimal.

How to use the workbook

You have probably chosen this workbook because you, or a tutor, have identified the need for you to develop skills, knowledge and understanding in the 'handling data' area of maths.

This workbook is divided into 2 sections:

- data and statistical measures

- probability.

Each section is divided into topics. Each topic begins with a short introduction, which puts the subject into context. Key terms are printed in **bold**. You can look up what these terms mean in the Glossary section at the end of the workbook.

The text is full of worked examples for you to follow. You can then put what you learn into practice by tackling the activities that follow.

There are suggestions for answers to each of the activities. Always check them before going on to the next activity. You may need to cover up the answers while working on an activity.

If you find that you made errors in the calculations, or got several answers wrong, go back over the relevant material in the section before attempting to move on to the next topic.

Section 1: Using data and statistical measures

Section 1: Using data and statistical measures

This section deals with handling data. But first of all, what does 'handling data' mean?

A dictionary gives the following meanings:

> hinders, race, contest in which chances are equalised by starts
> **handle** to use, operate, manage
> **handsome** of fine appearance, generouse, ample **handsomely** adv

> **dash** smash, throw, thrust, send with violence, cast down, tinge, flavour
> **data** facts, especially numerical facts, collected together for reference or information
> **date** day of the month, time of occurence

So, the phrase 'handling data' means being able to read, understand and interpret facts and figures.

We do this every day when we look at

- bus and train timetables;
- diagrams;
- charts; and
- graphs.

Items such as these are ways of showing complex information as simply as possible.

In fact, we are surrounded by mountains of data. This section will show you how to understand and analyse such data.

Section 1: Using data and statistical measures   HANDLING DATA

Reasons for completing this section

Working through this section will enable you to:

- read and understand information in the form of tables, diagrams, charts and line graphs

- collect, organise and show data in the form of tables, diagrams, charts and line graphs

- find the average in the form of an arithmetic mean for a set of data

- find the range for a set of data.

Resources

Resources you will need:

- A pen

- Some paper

- Some graph paper (or squared paper)

- A ruler

- A protractor (for measuring angles)

- A pair of compasses (or something to draw circles with)

- A holiday brochure or newspaper

- A map.

Topic 1: Understanding data

This topic will explain how to understand data in the form of:

- tables
- diagrams
- charts
- line graphs.

Take, for example, a holiday brochure. This is full of data which we need to understand. A holiday brochure uses:

- **tables** to show price lists
- maps or **diagrams** to show where the resort is or the distance to the airport
- **charts** and **graphs** to show temperatures and hours of sunshine.

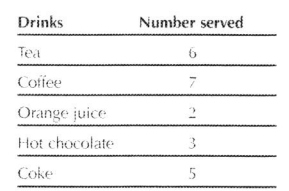

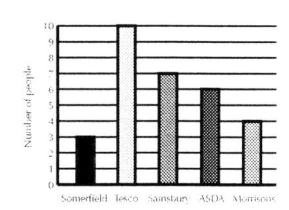

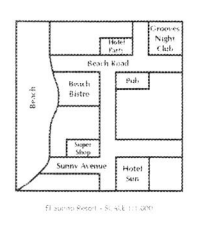

The brochure may provide all the information you need to compare holidays and pick the one you want. If you can, look through a holiday brochure and see for yourself. There is so much information and using tables, charts, graphs and diagrams makes it easier to understand.

You can see from things such as holiday brochures, newspapers and the news on television that there are different ways of showing information.

There are basically four ways of handling data to look at:

- tables
- diagrams
- charts
- graphs.

In this topic, you will be looking at each one in turn and finding out how to understand the data that is being shown.

Tables

Tables organise data to make it easier to understand. You can find tables all around:

- price lists
- timetables
- opening times on shops
- appointment cards.

Example 1

If you look in a newspaper somewhere it will probably tell you the weather forecast. It might even have it in a table like this:

The weather

Location	Today	Tomorrow
South and southwest	Sunny	Sunny with showers
Midlands	Sunny	Sunny
Scotland	Sunny	Storms
Wales	Showers	Storms
Ireland	Storms	Sunny

This could have been written out like this:

'The weather today in the south, southwest, Midlands and Scotland will be sunny. In Wales it will be showers and in Ireland there will be storms. Tomorrow it will be sunny with showers in the south and southwest. It will be sunny in the Midlands and Ireland and there will be storms in Scotland and Wales.'

Can you see how displaying it in table form made the information easier to understand?

Look back at the weather table. How is it drawn up?

Tables are made up of **rows** and **columns**.

The rows are horizontal – the bits that go across the page and the columns are vertical and go up and down.

Row
Row
Row

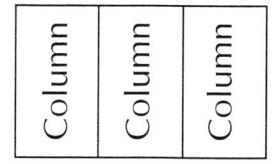

Column	Column	Column

To make sense of a table you need to have three things:

a) <u>A title</u>. This tells you what the table is about. In this table the title is 'The weather'.

b) <u>Row headings</u>. These tell you what is in each row. In our weather table the row headings are:

- South and southwest
- Midlands
- Scotland
- Wales
- Ireland

c) <u>Column headings</u>. These tell you what is in each column. In our weather table the column headings are:

- Location
- Today
- Tomorrow.

Tables can be very big with many rows and columns – it depends on how much information you are displaying.

For example, in a bus or train station you will see a huge timetable on the wall with many rows and columns. It is supposed to make the data easier to understand but it is still complicated and easy to get confused.

Section 1: Using data and statistical measures ■ Topic 1 HANDLING DATA

Activity 1

The local library has the following opening times:

Monday	9:30–12:30
Tuesday	12:30–5:30
Wednesday	9:30–5:30
Thursday	9:30–12:30
Friday	9:30–5:30
Saturday	9:30–12:30
Sunday	Closed

a) On what days is the library open all day?

b) What day is the library open only in the afternoon?

Check with the following suggestions before carrying on.

Suggestions for Activity 1

a) Wednesday and Friday

b) Tuesday

Now try the following self-check exercise.

Self-check 1

At the end of his shift a waiter drew up the following table to work out how many drinks he had served.

Drinks	Number served
Tea	6
Coffee	7
Orange juice	2
Hot chocolate	3
Coke	5

a) The table does not have a title. Make up a title you think suitable.

b) What are the row headings?

c) What are the column headings?

d) How many cokes did the waiter serve?

e) How many cold drinks did the waiter serve?

f) How many drinks did the waiter serve all together?

Answers to Self-check 1 are at the end of the section.

Diagrams

Scale drawings

Have you ever drawn a plan of your garden or a room in your house?

This is similar to scale drawings – the important thing with scale drawings, though, is that everything must be drawn to **scale**.

Section 1: Using data and statistical measures ■ Topic 1   HANDLING DATA

This means that everything must in proportion – everything must be 'shrunk' by the same amount.

All scale drawings **must** have a scale. This tells us how much the drawing has been shrunk.

Here is an example of a typical **scale drawing**.

Example 2

Note
This scale drawing has been drawn on squared paper. This makes it easier to draw and understand. Each square is 1 cm wide by 1cm long. So, instead of using a ruler you can just count the squares and this will tell you the measurement in centimetres.

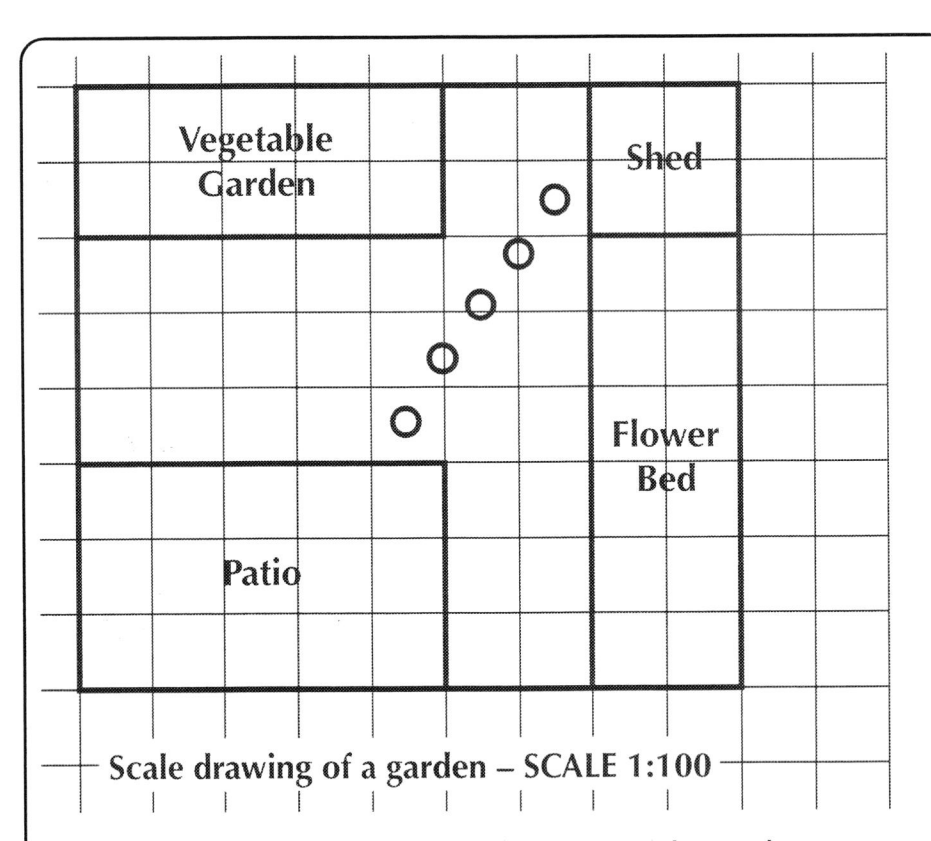

Scale drawing of a garden – SCALE 1:100

The scale drawing shows a garden. Using the scale, we can work out the sizes of the different parts of the garden.

The scale in this case is 1 : 100. This means that 1 cm on the scale drawing is equal to 100 cm in real life. 100 cm makes 1 m so here 1 centimetre on the drawing is 1 metre in real life.

- 1 : 100 means 1 cm on the drawing is equal to 1 m in real life.

Once we know the scale we can then go ahead and measure the distances on the drawing.

Using a ruler or just counting the squares we find that the patio is 5 cm wide and 3 cm long. This means that in **real life** it is 5 metres wide and 3 metres long.

HANDLING DATA   Section 1: Using data and statistical measures ■ Topic 1

> **Method summary**
> - Find out what the scale on the drawing is.
> - Measure the actual distance on the drawing using a ruler.
> - Multiply the distance you measure by the scale to give the distance in real life.

> **Note**
> Using the fact that 100 cm = 1 m and
> 1000 m = 1 km can help you change the units to make them easier to understand.

Now try this.

Activity 2

Using the scale drawing for the garden in the example above, find the following.

a) The width and length of the vegetable garden.

b) The width and length of the flower bed.

c) The distance from the patio to the vegetable garden.

d) I want to fit a trampoline that is 3 metres by 3 metres between the patio and the vegetable garden. Is there enough space for it?

Check with our suggestions below before continuing.

Suggestions for Activity 2

a) 5 m wide by 2 m long

b) 2 m wide by 6 m long

c) 3 m

d) With the distance between the patio and vegetable garden only being 3 metres and with the trampoline being 3 metres wide it would be a bit of a squeeze.

Maps

Maps are very similar to scale drawings. The main difference is that they are usually used to show places.

If you look in a holiday brochure you will see lots of maps.

They are often used to show how a resort is laid out. They show where only a few, important places are, e.g. shops, hotels, beach, swimming pools and restaurants.

It is important to understand how to read the map so that you do not end up somewhere that is too far from the facilities you want to be near/avoid.

Example 3

Here is a typical example of a map you might find in a holiday brochure.

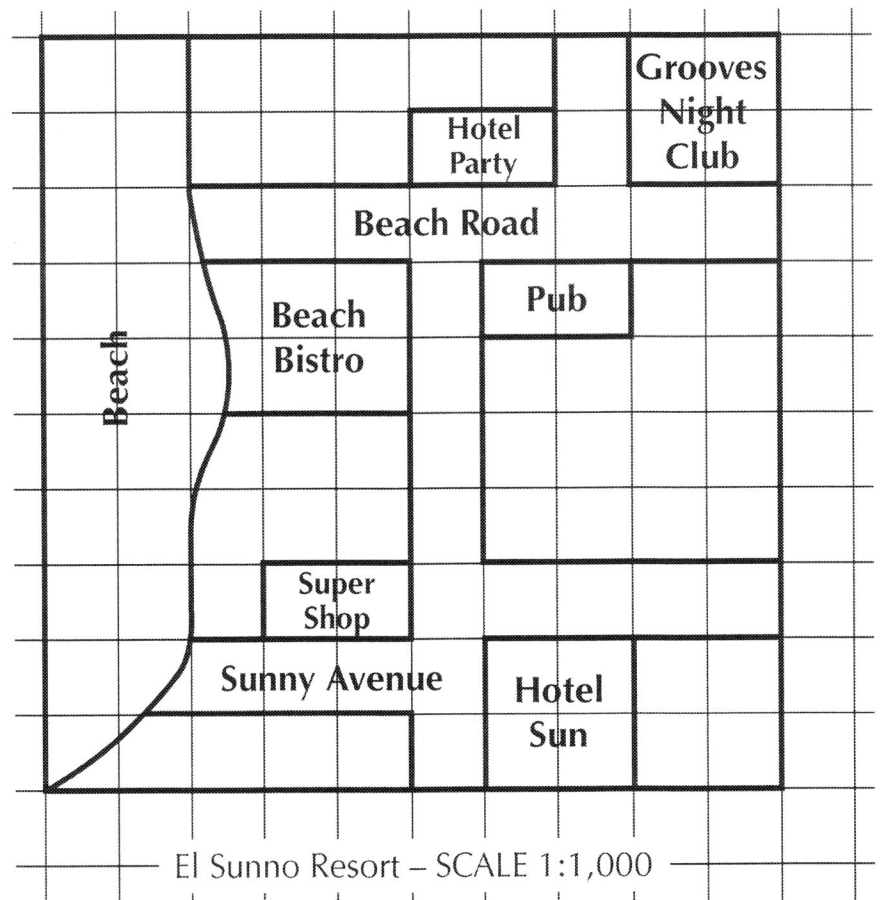

Note
The entrances to the buildings are marked with crosses on this map. You need to measure to and from these crosses.

El Sunno Resort – SCALE 1:1,000

As with scale drawings the thing you need to know before you can understand the map is the **scale**.

HANDLING DATA   Section 1: Using data and statistical measures ■ Topic 1

Example 3 continued

The scale basically tells you how much the map has been 'shrunk'.

In this case the scale is 1 : 1,000. This means that every 1 cm on the map represents 1,000 cm or 10 m in real life.

Using the scale we can interpret the data on the map. We can work out how far different places are from one another.

To do this we need to measure the distances on the map and then multiply them by 1,000 to get the actual distance in centimetres. Or, more simply, multiply by 10 to get the actual distance in metres.

So, in the case of our holiday resort the nightclub is 1 cm from 'Hotel Party', this multiplied by 10 makes the real distance 10 metres – not very far at all. This could then affect whether you choose to stay at Hotel Party (depending on whether you like nightclubs or not).

Remember
To multiply by 10 all you need to do is add a zero at the end of the number you are multiplying, so 7 x 10 = 70 and 30 x 10 = 300.

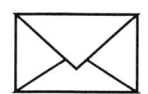

Investigation

If you can, have a look at a local road or street map. Can you find the scale? Use the scale to work out distances to local places from where you live.

Note
Using the fact that 100 cm = 1 m and 1,000 m = 1 km you might like to change the units to make them easier to understand.

Method summary

- Find out what the scale on the map is.

- Measure the actual distance on the map using a ruler.

- Multiply the distance you measure by the scale to give the distance in real life.

© National Extension College Trust Ltd   DEVELOP YOUR MATHS

Section 1: Using data and statistical measures ■ Topic 1   HANDLING DATA

Self-check 2

Using the map for the holiday resort in Example 3, find the following distances.

a) The distance in real life from the Waves pub to Hotel Sun.

b) The distance in real life from the shop to the Beach Bistro.

c) The distance in real life from the nightclub to the beach.

Answers to Self-check 2 are at the end of this section.

Pictograms

Pictograms use pictures to count with. They are a simple way of showing data.

Example 4

The following pictogram is showing the number of cars using a car wash at different times during the week.

Sunday afternoon   🚗 🚗 🚗

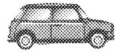

Monday morning   🚗

KEY: 🚗 means one car using the car wash

The important thing to remember with pictograms is that there must be a **key** to tell the reader what the picture means. In Example 4 the picture of one car means one car used the car wash but in the next example the key is used differently – look at the following example and see if you can work out how.

HANDLING DATA   Section 1: Using data and statistical measures ■ Topic 1

Example 5

People buying petrol from a garage

Sunday afternoon ☺

Monday morning ☺ ☺ ☺

You might think the key is

☺ means one person buying petrol

but here

☺ means 5 people buying petrol

☺ means 4 people

and

☺ means 3 people

Activity 3

Can you work out what ☺ and ☺ mean?

Check with our suggestion, then move on.

© National Extension College Trust Ltd   DEVELOP YOUR MATHS

Suggestion for Activity 3

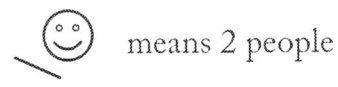

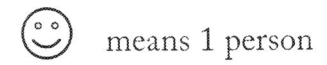

☺ means 2 people

☺ means 1 person

So, the key can be used to show more than one item. This could be done to make the drawing of the pictogram easier when working with bigger numbers.

It is important to make sure you understand what the key means so that you can understand the data correctly.

There are good and bad points to using pictograms:

- The good thing with using pictograms is that they are easy to understand.

- The problem with using pictograms is that they can only show a few things.

HANDLING DATA   Section 1: Using data and statistical measures ■ Topic 1

Self-check 3

Here is a pictogram showing the number of people who visited the dentist over 5 days. Count up the symbols and fill in the table below and then see if you can answer the following questions.

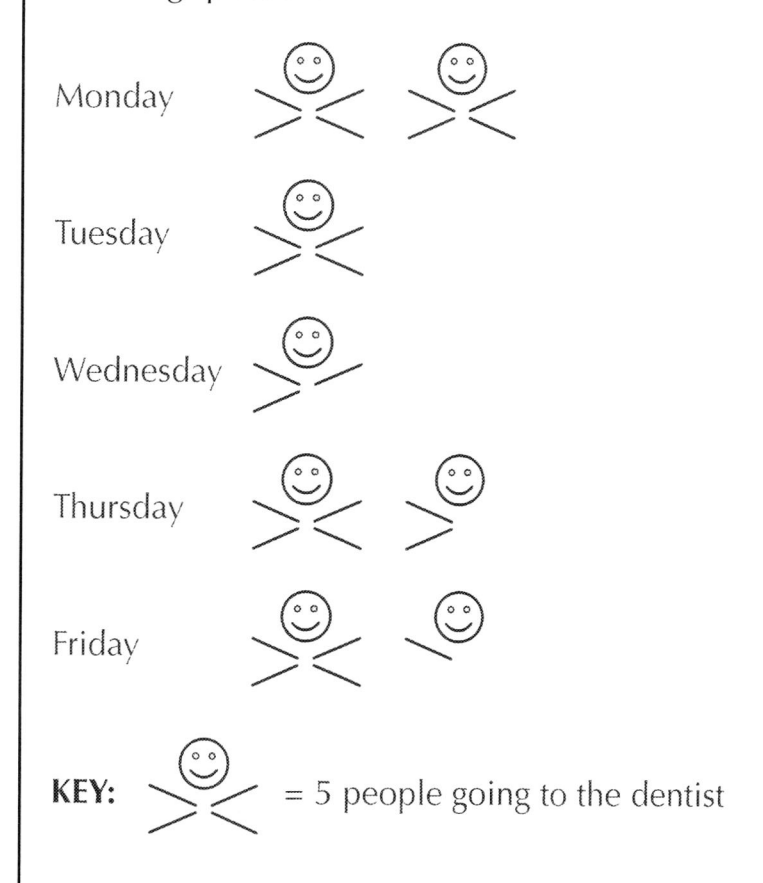

Monday   ☺ ☺

Tuesday   ☺

Wednesday ☺

Thursday ☺ ☺

Friday   ☺ ☺

KEY: ☺ = 5 people going to the dentist

Monday	Tuesday	Wednesday	Thursday	Friday

a) On what day did most people visit the dentist?

b) On what day did the least people visit the dentist?

Answers to Self-check 3 are at the end of the section.

Charts

Charts are basically maths pictures. There are two types of charts: pie charts and bar charts.

Pie charts

A **pie chart** is a big circle (or 'pie') cut into pieces. Here is an example:

> **Note**
> Pie charts must contain both a title and a key explaining what each segment/colour means.

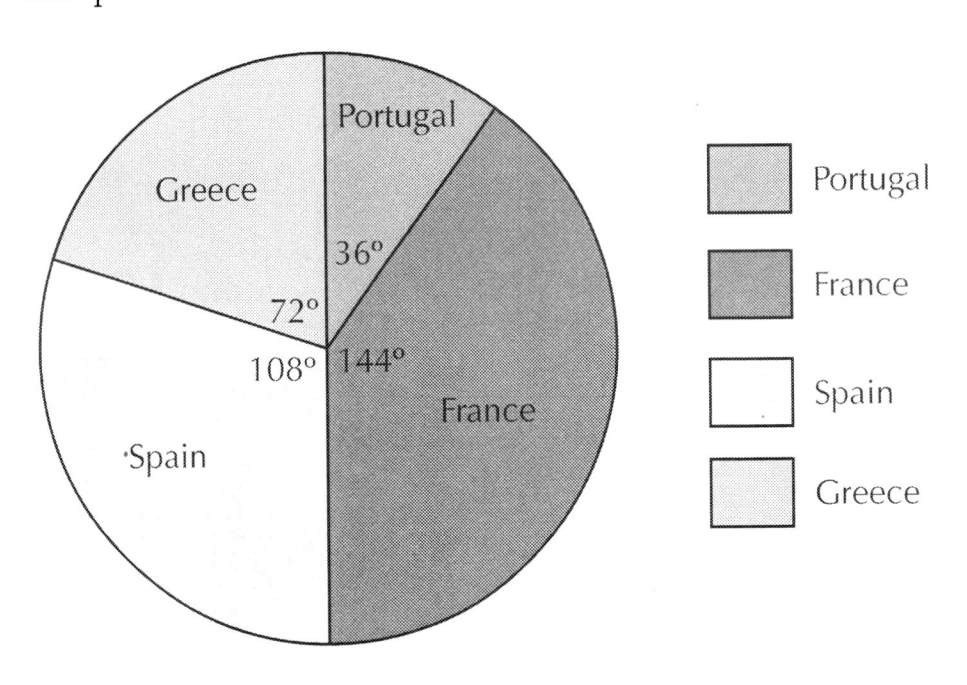

Pie chart showing the top holiday destinations for a group of workers in a factory

Q. When do we use pie charts?

A. Usually when we have a few bits of information to show. It is a clear way of showing data because you can quickly see what has the biggest or smallest 'slice' of the 'pie'.

From the above pie chart you can see that the top holiday destination was France because this has the biggest slice of 'pie'.

You are also given the size of each segment (in degrees) and so if you know the amount of people the chart is dealing with you can work out how many people went to Spain, Portugal, France and Greece.

HANDLING DATA Section 1: Using data and statistical measures ■ Topic 1

> **Remember**
> A circle is divided into 360 degrees (usually written as 360°). A semi-circle (or half a circle) is thus 180°, a quarter of a circle is 90° and so on.

> **Note**
> Don't be confused by dividing by 360. Use a calculator if you want. The important thing is to understand the METHOD being used.

> **Remember**
> This is maths in real life. Look at your answer. Does it make sense? If my answer was 53.5 then my common sense would tell me you can't have half a person. So, I'd have another look at my working.

Let's say the survey was dealing with 180 people. Using the fact that the whole 'pie' is 360° and this equals 180, then you have

360° = 180 people

and thus

1° = 180 ÷ 360 = 0.5 people

So for Spain with 108° of the circle or 'pie' you can see that

108° = 108 x 1°= 108 x 0.5 people = 54 people

so 54 people from the factory went to Spain.

If you are not told the angles you can measure them using a protractor. This is a bit like a ruler but measures angles not distances. It is a see-through semi-circular piece of plastic and looks like this

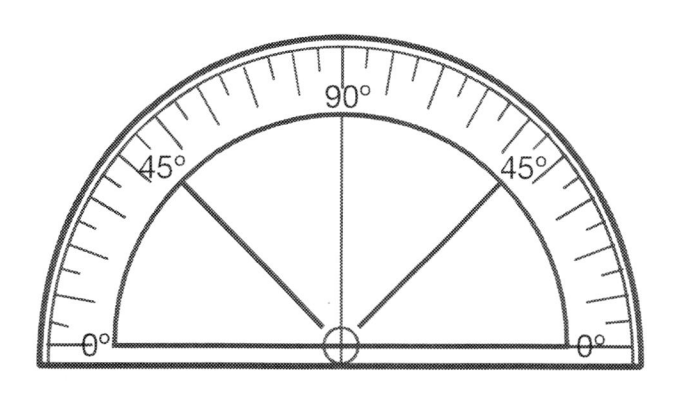

© National Extension College Trust Ltd   DEVELOP YOUR MATHS   21

Section 1: Using data and statistical measures ■ Topic 1   HANDLING DATA

- To use a protractor you place it on top of the angle you are trying to measure.

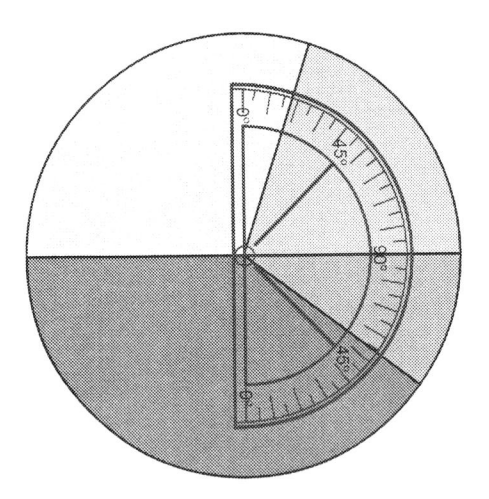

- You then line up one of the lines of your angle with one of the lines at the edge of the protractor (it will have 0° written on it).

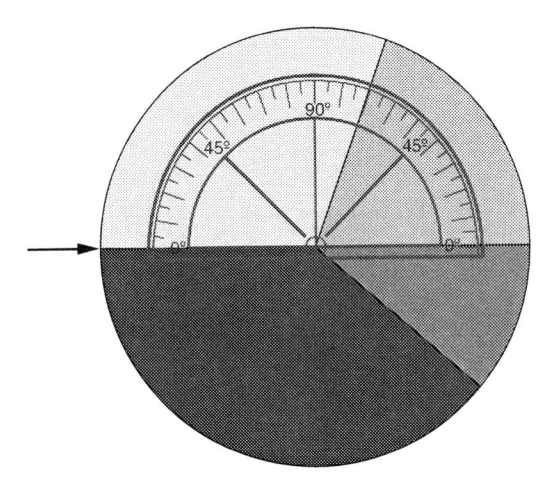

- Then line up the other line of your angle and read off the angle.

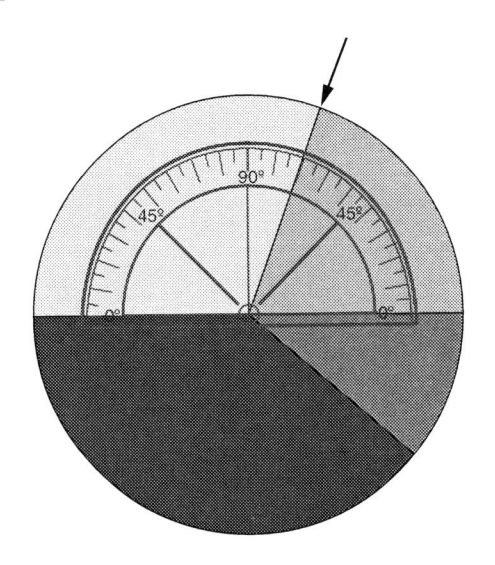

HANDLING DATA   Section 1: Using data and statistical measures ■ Topic 1

> **Method summary**
> - Find out what the whole of the 'pie' represents.
> - Divide this by 360 to give 1° (use a calculator if you need to).
> - Multiply the amount for 1° by the size of each segment.
> - This gives what each segment actually represents.
> - If you have not been given the size of the segments (in degrees) then you measure them using a protractor and go from there.

Now try this activity.

Activity 4

> Go back to the pie chart and work out the number of people who went on holiday to:
> a) Portugal
> b) Greece
> c) France.

Check with the following suggestions before continuing.

Suggestions for Activity 4

a) 18 people

b) 36 people

c) 72 people

Section 1: Using data and statistical measures ■ Topic 1   HANDLING DATA

Bar charts

A **bar chart** is a different type of chart.

- It uses boxes or bars to show the data.
- The axis on the bottom (horizontal axis) tells you what each bar is.
- The axis going up the side (vertical axis) tells you how many 'things' that bar relates.

Example 7

The following table shows how 20 people paid for their goods in a shop.

Method of payment	Number of people
Cash	5
Cheque	1
Switch	9
Storecard	2
Visa	3

This could be shown in a bar chart as follows:

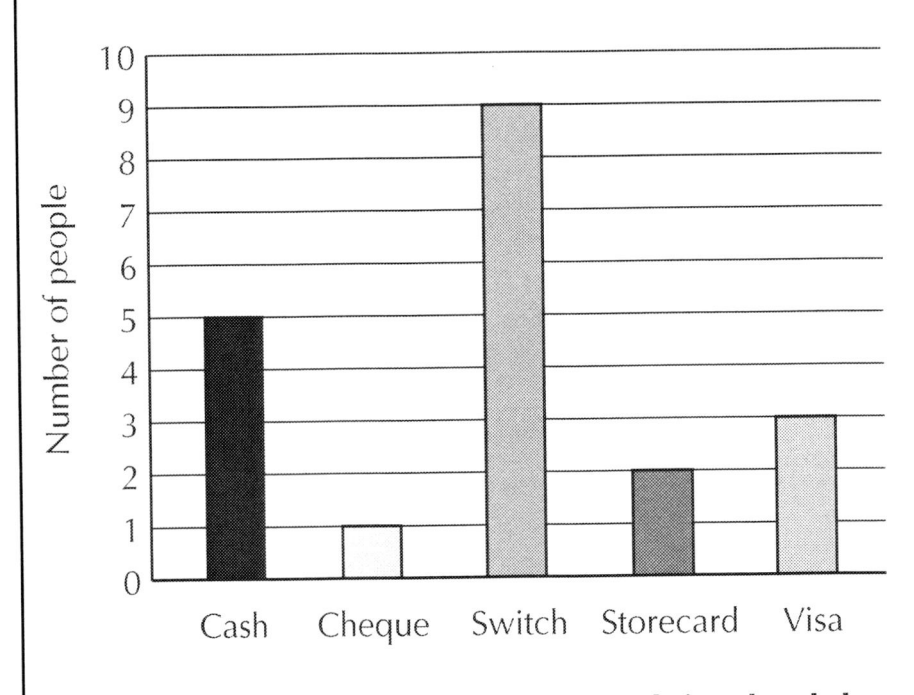

How 20 people paid for their goods in a local shop

HANDLING DATA   Section 1: Using data and statistical measures ■ Topic 1

It is easy to see from the chart that most people paid for their goods using Switch.

> **Remember**
> To read the numbers on a bar chart all you need to do is line up the top of the bar you are interested in with the axis going up the side and read off the number.

Now try the following activity.

Activity 5

1. What do you think the advantages of using bar charts are?
2. What disadvantages are there?

Check with the following before continuing.

Suggestions for Activity 5

a) The advantage of using bar charts is that it is easy to compare the sizes of the bars to see trends.

b) The disadvantage of using bar charts is that only simple information can be shown.

Section 1: Using data and statistical measures ■ Topic 1 HANDLING DATA

Self-check 4

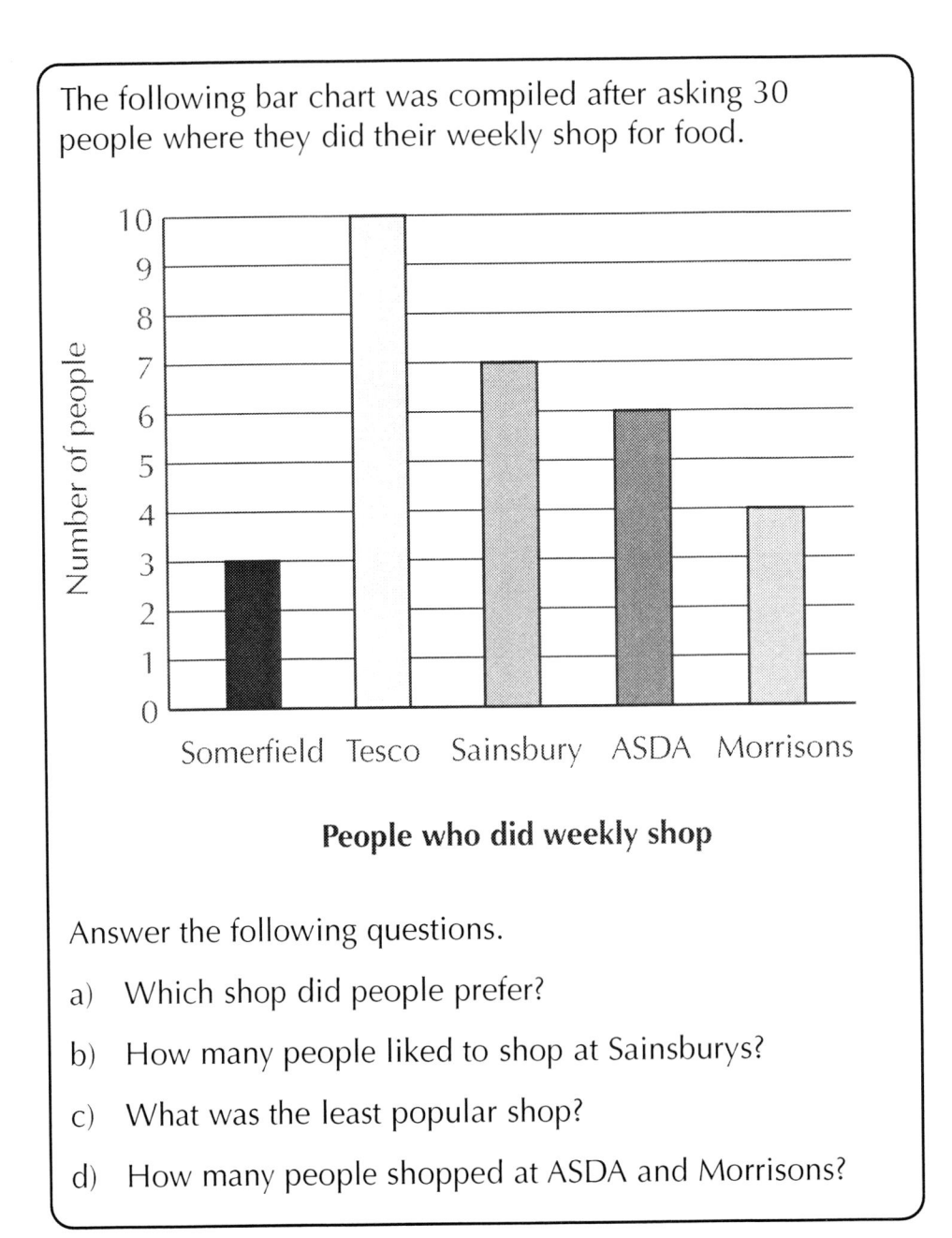

The following bar chart was compiled after asking 30 people where they did their weekly shop for food.

People who did weekly shop

Answer the following questions.

a) Which shop did people prefer?

b) How many people liked to shop at Sainsburys?

c) What was the least popular shop?

d) How many people shopped at ASDA and Morrisons?

Answers to Self-check 4 are at the end of the section.

Graphs

Line graphs are drawn by plotting points and then joining them with straight lines. Line graphs are used because they make it easy to compare data and see trends.

HANDLING DATA   Section 1: Using data and statistical measures ■ Topic 1

Example 8

The number of microwaves sold at an electrical shop is shown in the table below.

Month	Oct.	Nov.	Dec.	Jan.	Feb.
Sales	2	4	10	4	1

This data can be displayed in a line graph as follows.

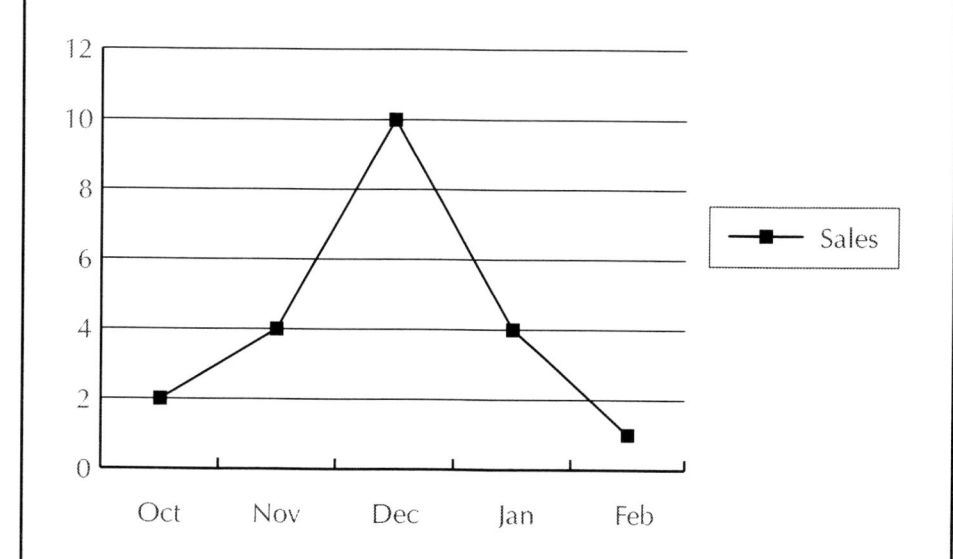

[Line graph showing Sales by month: Oct=2, Nov=4, Dec=10, Jan=4, Feb=1, with y-axis from 0 to 12]

You can see from the graph that sales peaked during December. This was to be expected because of Christmas but the line graph shows this very clearly.

This is the advantage of line graphs – you can see trends, peaks and troughs clearly.

The problem with them is that only simple information can be shown.

Now try the following activity.

Section 1: Using data and statistical measures ■ Topic 1   HANDLING DATA

Activity 6

In hospitals, line graphs are often used in patients' records. One way they are used is to display temperature records to monitor the patient's progress. Here is a line graph showing the temperature of a patient taken over 4 hours. Use it to answer the questions below.

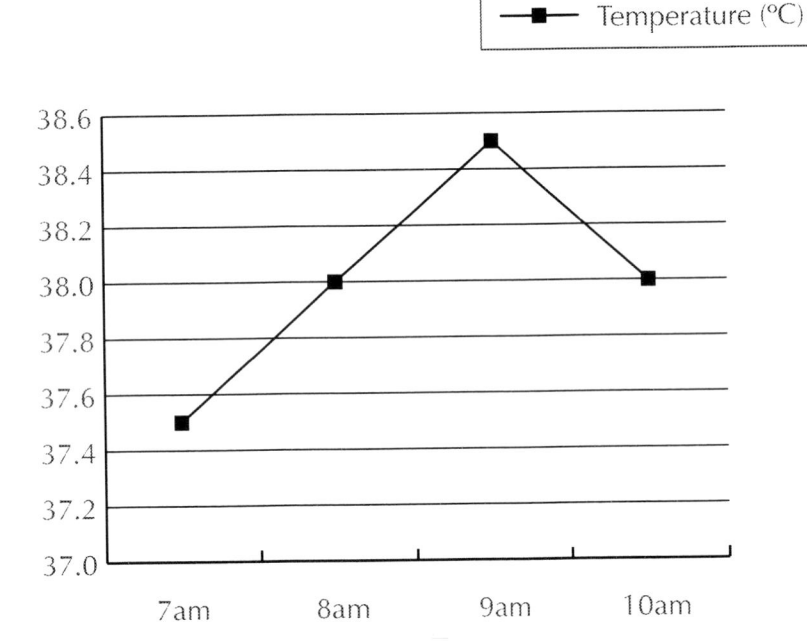

a) What was the highest temperature recorded?

b) What was the lowest temperature recorded?

c) Do you think the patient's temperature will go up or down at 11 a.m.?

Check with the following before continuing.

Suggestions for Activity 6

a) 38.5°C (at 9 a.m.)

b) 37.5°C (at 7 a.m.)

c) It looks as if the patient's temperature will continue to go down.

HANDLING DATA   Section 1: Using data and statistical measures ■ Topic 1

Self-check 5

> Before moving on to the next topic look back over the unit. We've looked at tables, diagrams, charts and graphs. Do you feel confident about understanding the data displayed in these different ways? If not, then have another look at this topic. If you feel happy with understanding data, then answer the following questions.
>
> a) What must a pictogram have?
>
> b) What is the problem with using pie charts?
>
> c) What does a scale tell you?
>
> d) Why are titles important for tables, diagrams, charts and graphs?

Answers to Self-check 5 are at the end of the section.

Summary

In this topic you have:

- learned how to read and understand information shown in tables, diagrams, charts and graphs

- understood that the titles, labels, keys and scales provide information

- learned how to read the scale on a diagram or map

- learned how to use a simple scale such as 1 cm to 1 m.

> By completing this topic you have covered HD1/L1.1 of the Data Handling section of the curriculum:
>
> - Extract and interpret information (e.g. in tables, diagrams, charts and line graphs).

Topic 2: Collecting and showing data

In this topic you will learn how to collect, organise and show data in tables, diagrams, charts and line graphs.

In Topic 1 you looked at how different types of data can be displayed to help us understand it quicker and easier using

- tables
- diagrams
- charts
- line graphs.

In this topic you will be building on this by

- collecting
- organising and
- showing data.

We will be using the same ways of doing this – using tables, diagrams, charts and line graphs.

Another word that means 'collecting information' is a survey.

Have you ever been stopped by someone doing a survey or had to fill in a survey? People try to find out things like:

- what we like
- what we do
- what we buy
- what we watch on television
- what we eat.

When they do this it is **collecting data.**

But that is not the end of it.

This information then needs to be organised and displayed so that it is easy to understand. We will be doing this in this topic for **discrete data.**

Discrete data is data made up of things that are separate and can be counted.

Some examples of discrete data are:

- the number of people on a bus
- the number of cars in a car park and
- the number of leaves on a tree.

All these things are separate items that we can count (if we want to).

Collecting data

A **tally chart** is a useful way of collecting information. A tally chart consists of a series of tallies.

Tallying works like this:

- for each unit you make a tally mark like this: |
- when you have five units you 'cross out' the other four tally marks like this: ⊬⊦⊦
- you use groups of fives as follows:

    |||| = 4

    ⊬⊦⊦ = 5

    ⊬⊦⊦ | = 6

    ⊬⊦⊦ ⊬⊦⊦ = 10

Activity 7

Write the following numbers in tally form:

a) 3

b) 7

c) 9

d) 14

e) 18

HANDLING DATA

Section 1: Using data and statistical measures ■ Topic 2

Check with the following before continuing.

Suggestions for Activity 7

a) |||

b) ||||| ||

c) ||||| ||||

d) ||||| ||||| ||||

e) ||||| ||||| ||||| |||

> **Note**
> Tally charts and tally tables are the same thing.

We can use tally charts to record data when we carry out surveys and collect data.

Have you ever seen people parked at the side of a road doing a traffic survey? They could be recording the number of people in each car and the tally chart would look something like this:

No. of people in car	No. of cars	Total
1		
2		
3		
4		
5+		

As the cars go by they record the information like this:

No. of people in car	No. of cars	Total
1	||||	
2	|||	
3	|	
4	||	
5+	|	

© National Extension College Trust Ltd DEVELOP YOUR MATHS

At the end of the survey they would then add up the tallies in the middle column and put the totals in the right-hand column.

Q. Why do we use tally charts?

A. It is a quick and simple way of recording data.

Self-check 6

Tip
Tick or cross off each entry as you put it into your tally chart. This will help you stop losing your place.

In a survey 20 people were asked how many people lived in their house. These were the answers:

1	2	0	3	2
1	2	1	0	1
0	2	3	1	1
2	0	2	1	1

Make a tally chart of this data.

Answers to self-checks are at the end of the section.

Investigation

Carry out your own survey. Ask at least 10 people what their favourite types of TV programmes are. Your categories should be

- soaps
- dramas
- nature programmes
- documentaries
- none of the above.

Organising data

Once we have our data the next step is to organise it.

We organise data to make it easier to understand.

We will be looking at organising data in the same four ways we looked at in Topic 1:

- tables
- diagrams
- charts
- line graphs.

Tables

So, we have our data. Now we want to display it to make it as easy to read as possible.

The first way we will look at doing this is with **tables**.

Tables are made up of rows and columns and need to have three things to make them clear:

- a title explaining what the table is about
- row headings telling you what the rows mean
- column headings telling you what the columns mean.

These three things are what you need to decide before you start your table.

If you have a tally chart then you probably have headings for your rows and columns already and you just need to decide on what your title for the table will be. To do this just think what the data is about. What is it showing?

The more details you put in your title the better. Ideally, you need to give the following information (if you have it).

- What the data is.
- Where it was recorded.
- When it was recorded.

Let's go back to our traffic survey. The tally chart looked like this:

No. of people in car	No. of cars	Total
1	IIII	4
2	III	3
3	I	1
4	II	2
5+	I	1

We could do this in table form in two ways:

EITHER

No. of people in car	1	2	3	4	5+
No. of cars	4	3	1	2	1

OR

No. of people in car	No. of cars
1	4
2	3
3	1
4	2
5+	1

Either way would be OK.

Which way do you think is clearest?

What title do you think the table should have?

> **Note**
> Ideally you would want to have the date of the survey and exactly where it was carried out.

HANDLING DATA Section 1: Using data and statistical measures ■ Topic 2

You could have something like:

'A traffic survey showing the number of people in cars on a local road on one day of the week.'

Activity 8

> The following information shows the colours of cars in a car park one lunchtime. Draw a table to present the data.
>
> | | | | | |
> |---|---|---|---|---|
> | red | yellow | red | blue | white |
> | blue | black | white | red | green |
> | red | white | green | black | blue |
> | white | blue | red | red | black |

Check with the following before continuing.

Suggestions for Activity 8

Number of cars with certain colours in a car park one lunchtime

Colour of car	Number of cars
Red	6
Yellow	1
Blue	4
White	4
Black	3
Green	2

Diagrams

Scale drawings and **maps** are really simple plans of a space or place.

- The first thing you need to decide when drawing a scale drawing or map is the scale – how much you are going to shrink the things you want to put in your diagram.

Section 1: Using data and statistical measures ■ Topic 2   HANDLING DATA

- You then need to measure the objects or distances in real life and divide these by your scale so that you can fit them on your diagram.

- You are then ready to draw your diagram. Squared paper (or graph paper) helps because it means you can use the squares as part of your scale and it makes things easier to draw.

- Don't forget to give your diagram a title and put the scale in.

Example 9

I want to re-design my garden.

Before starting I decide to draw a scale diagram to see how the garden looks now.

The garden is 8 metres wide and 12 metres long. There is a flower bed running down the left side which is 2 metres wide and a large shed in the bottom right corner that is 3 metres wide and 4 metres long. Draw a scale drawing showing the garden as it is.

- The first thing I need to do is decide on my scale. I have 1-cm squared paper and so the easiest thing to do is use the scale of 1 cm to 1 m. There are 100 cm to 1 m so this is a scale of 1 : 100.

- Now I draw the whole garden on my squared paper. The garden is rectangular so I can draw a box. It is 8 m wide so this is 8 squares across and 12 m long so that is 12 squares down.

- Next, I want to draw in the flower bed. This is down the left side and is 2 m wide. This means that I draw a line 2 squares out from the left side of the box. I label this box 'FLOWER BED'.

- Then, I want to draw in the shed. This is 3 m wide and 4 m long so I draw a box in the bottom right corner that is 3 squares out from the right side and 4 squares up. I label this box 'SHED'.

- Finally, I give my diagram a title and then write in the scale.

HANDLING DATA   Section 1: Using data and statistical measures ■ Topic 2

Example 9 continued

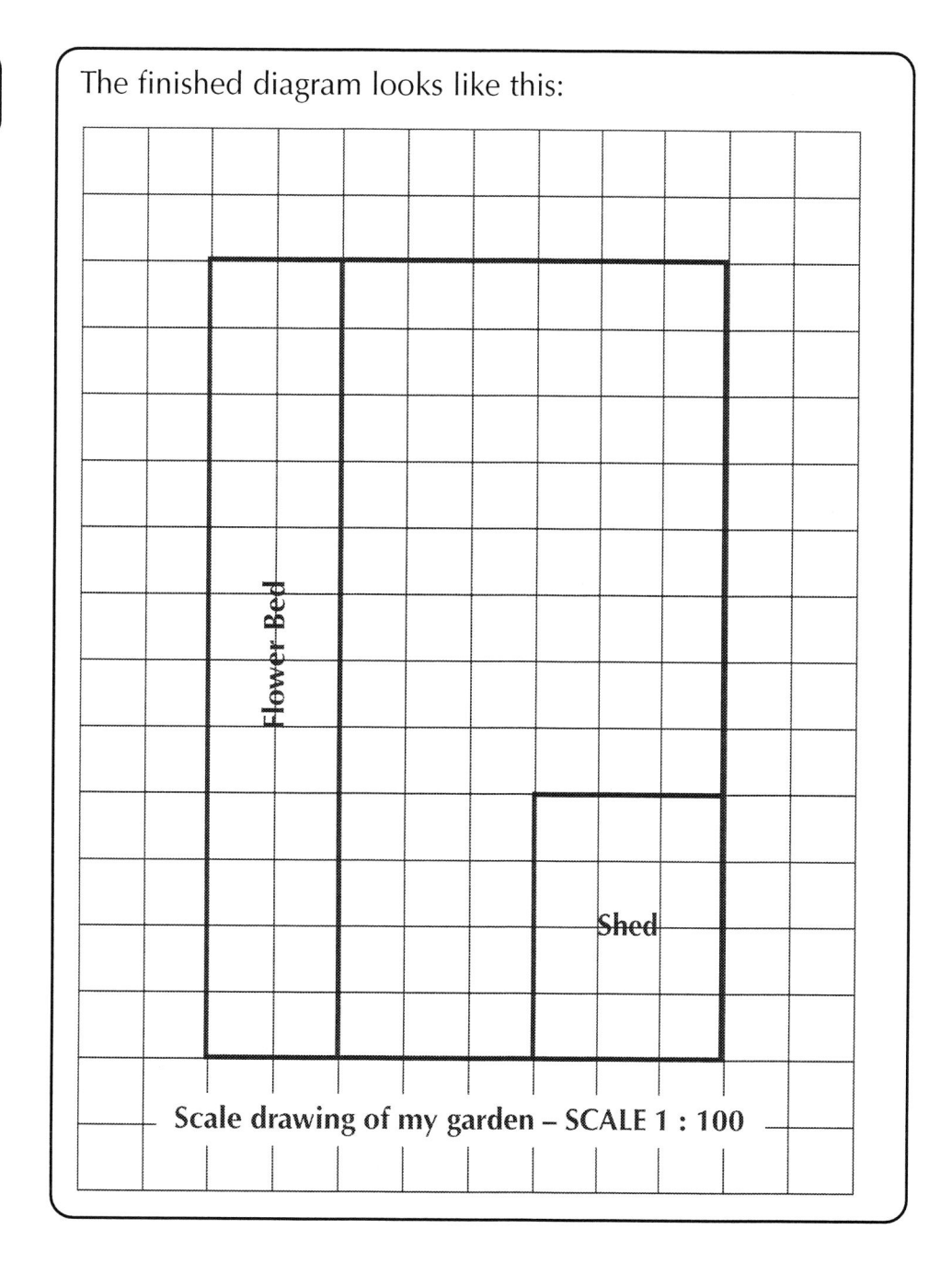

The finished diagram looks like this:

Flower Bed

Shed

Scale drawing of my garden – SCALE 1 : 100

© National Extension College Trust Ltd   DEVELOP YOUR MATHS   39

Section 1: Using data and statistical measures ■ Topic 2   HANDLING DATA

Self-check 7

After drawing out a scale drawing of my garden I start thinking about adding some features.

Using squared paper (if you have any) copy the previous scale diagram and add the following features.

a) I would like to have a patio so I can sit outside. The sunniest place is in the top right corner. I would like it to be 4 m wide and 4 m long. Draw this on the scale drawing.

b) Everyone seems to have a water feature in their gardens. I would like to have a pond at the top of the garden between the patio and the flower bed. The size of the pond should be 2 m wide and 1 m long. Draw this on the scale drawing.

Answers are at the end of the section.

Pictograms

Pictograms are types of tables that use symbols rather than numbers.

Pictograms can be a good way of showing information because they have a strong visual impact.

As with tables you need to decide on your title and what each row of the pictogram means. You also need to decide on your key. The key tells your reader what the picture you are using means.

As a reminder let's have another look at the pictogram we used in Topic 1.

People buying petrol from a garage

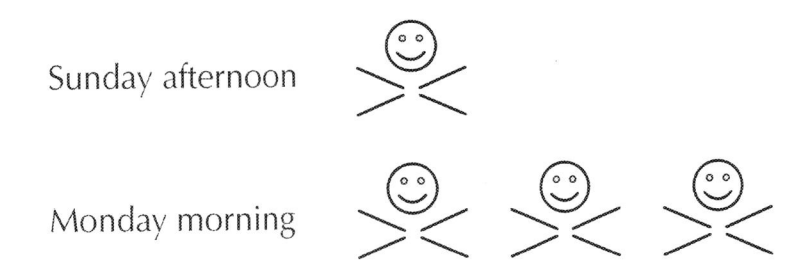

Sunday afternoon   ☺

Monday morning   ☺ ☺ ☺

Can you remember that

☺ meant five people buying petrol

and not one person as you might have thought? This is why it is important to put a key. It is no good drawing a pictogram if nobody else can understand it!

So, to draw a pictogram you need to do the following.

- Decide what key are you going to use. It is good to keep this as simple as possible (and something you can draw easily).
- Choose a title.
- Choose the row headings.

Activity 9

The following table shows the number of people queueing at a local post office at different times of the day. Show this information as a pictogram.

Time	Number
9 a.m.	4
11 a.m.	2
1 p.m.	7
3 p.m.	1
5 p.m.	3

Check with our suggestions before continuing.

Suggestions for Activity 9

People queueing at a local post office during different times of the day

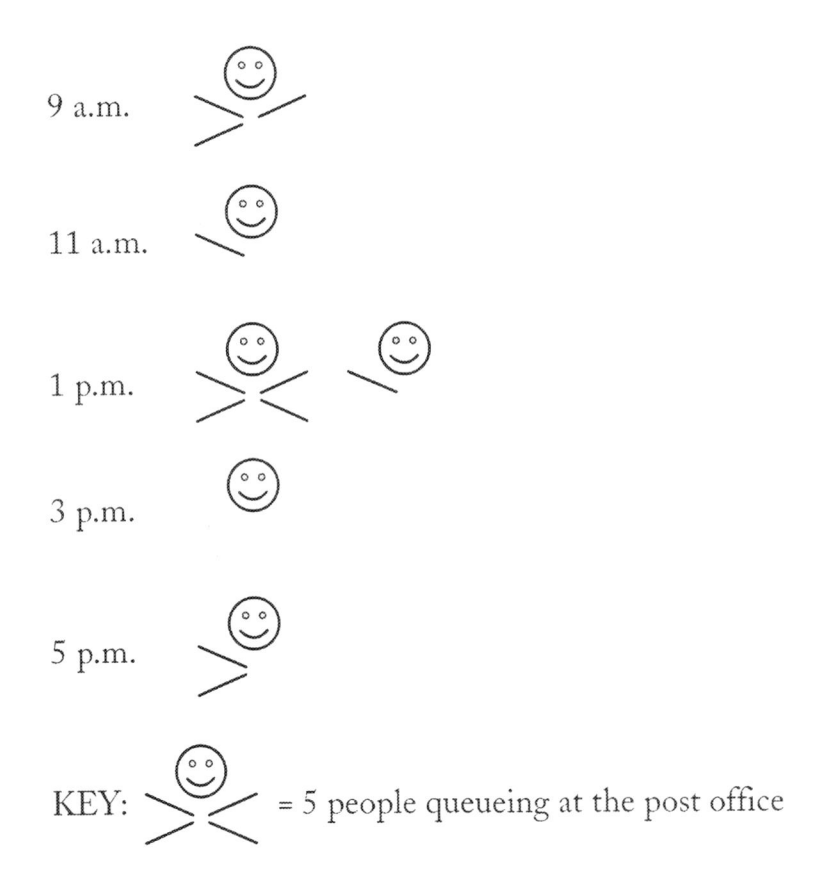

9 a.m.

11 a.m.

1 p.m.

3 p.m.

5 p.m.

KEY: = 5 people queueing at the post office

Pie charts

Pie charts are clear ways of presenting data but can be difficult to draw and the calculations can be complicated.

Pie charts are circles divided in sections (or slices). The sizes of these sections represent the data.

HANDLING DATA   Section 1: Using data and statistical measures ■ Topic 2

Example 10

In a survey 36 people were asked what their favourite soap opera was. Their responses were as follows:

Coronation Street   18

Eastenders   9

Hollyoaks   6

Other/none   3

To draw our pie chart we first need to work out the size of each 'slice' of our circle (or pie).

To do this we need to remember that angles are measured in degrees (written as °) and that a circle is divided into 360°.

In this example we are told that 36 people were surveyed and so we need to work out how many degrees of our circle one person makes.

So, we have 360° = 36 people

and thus

1 person = 360 ÷ 36 = 10°

So every 10° of the pie represents one person.

We can then work out what the size of each slice (or category) should be:

Soap opera	Number	Angle
Coronation Street	18	18 x 10° = 180°
Eastenders	9	9 x 10° = 90°
Hollyoaks	6	6 x 10° = 60°
Other/none	3	3 x 10° = 30°

We can now start to draw our pie chart.

First, we need to draw a circle. If you have a pair of compasses then that's the best way. However, drawing around a circular object or a double protractor will be fine but you do need to find out the centre of the circle (you might be able to guess).

Section 1: Using data and statistical measures ■ Topic 2   HANDLING DATA

Example 10 continued

Then draw a line from the centre of the circle to the top of the circle. This will be the line you start drawing your 'slices' from.

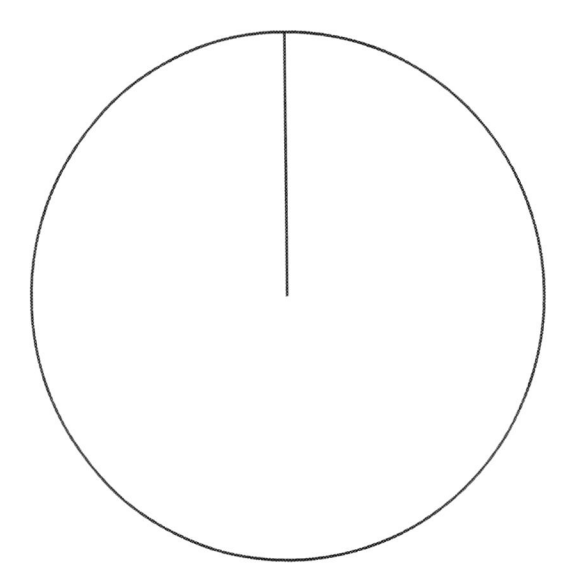

Using a protractor, measure each 'slice' one after the other.

To draw angles with a protractor you need to do the following:

- Place the protractor on top of the line you are drawing an angle from.

- Count along the protractor the number of degrees you want your angle to be.

- Mark this on your paper.

- Use the straight side of the protractor to draw a line from the line you started with to the mark you have just made.

HANDLING DATA   Section 1: Using data and statistical measures ■ Topic 2

Example 10 continued

In this example, you will draw a slice that is 180° (for Coronation Street) then from that line a slice of 90° for Eastenders. Work through all the data categories until you get to the line you started with.

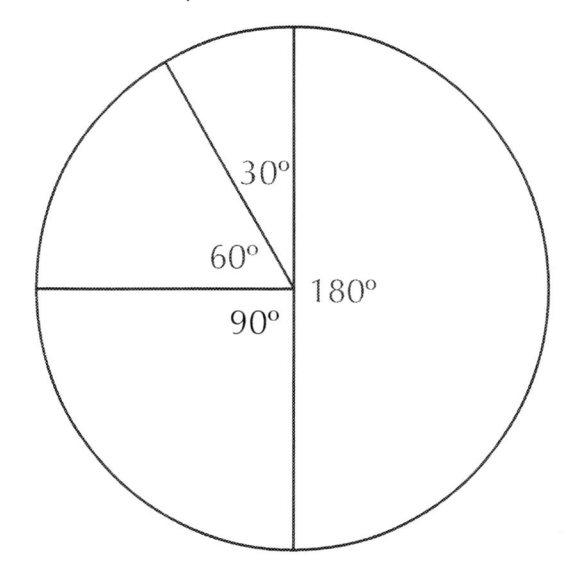

Label each slice (you can put the degrees in).

Give your pie chart a title.

Here is the pie chart for the soap opera data.

Favourite soap operas watched

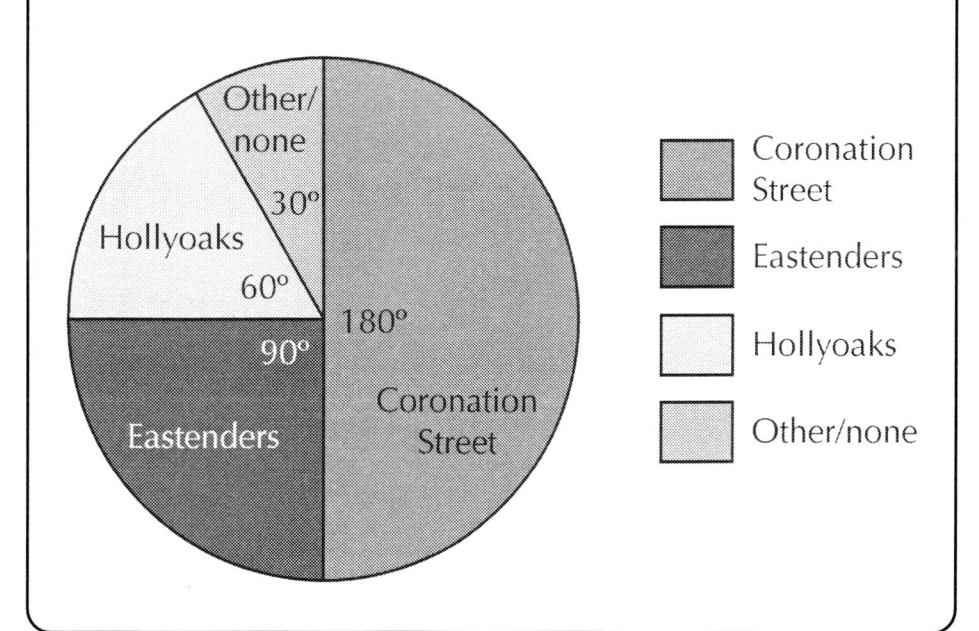

Section 1: Using data and statistical measures ■ Topic 2 HANDLING DATA

> **Method summary**
>
> - Find out what the whole of the 'pie' is going to represent (this is the total of your categories added together).
> - Divide 360 (the size of a circle in degrees) by this total to tell you what one unit (of your data) makes (use a calculator if you need to).
> - Multiply the amount for this one unit by the size of each category.
> - This gives the size of what each segment should be (in degrees).
> - Draw a circle and draw a line from the middle of the circle to the top of it.
> - Starting from this line and using a protractor measure and draw each segment.
> - Label the segments.
> - Give your pie chart a title.

Activity 10

In a survey, 18 people were asked what their favourite pets were. The data was as follows.

Pet	Cat	Dog	Rabbit	Bird	Fish
Frequency	5	6	4	1	2

Draw a pie chart to represent this information.

Check with the following before continuing.

Suggestions for Activity 10

Favourite pets

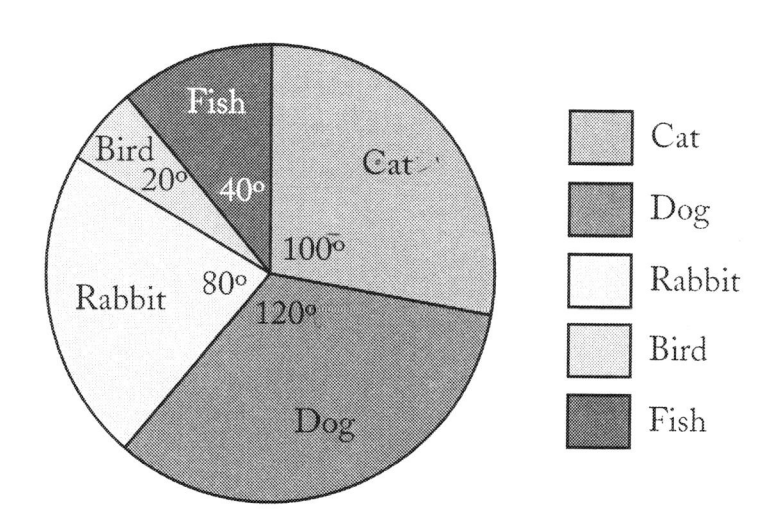

Bar charts

Let's go back to our traffic survey.

In table form the data was displayed as:

No. of people in car	No. of cars
1	4
2	3
3	1
4	2
5+	1

We can display this data in a **bar chart,** as follows.

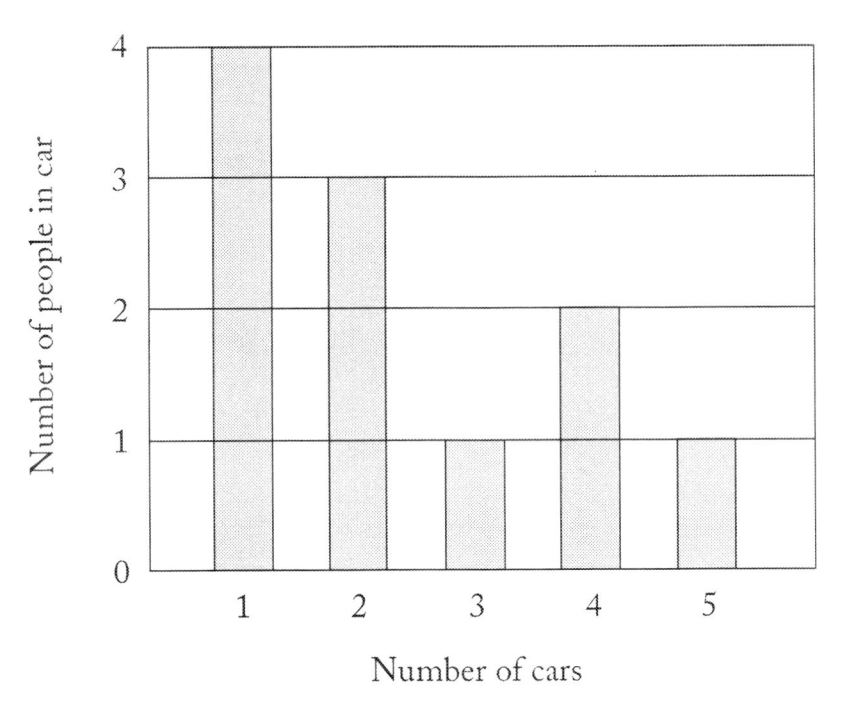

Number of people in cars travelling on a local road one morning

(Vertical axis: Number of people in car; Horizontal axis: Number of cars)

The good things about bar charts is that they show data clearly. To do this bar charts must contain the following information:

- A title explaining what the bar chart means.

- Labels telling you what each bar means. This could be a key or just a label underneath the line on the bottom (the horizontal axis).

- The line going up on the left side (the vertical axis) must have numbers at equal intervals. This is telling you how big the bars are so that your reader can read off the data.

You need to decide on your labels and what number intervals (i.e. how 'tall' your bars are going to be) you are going to use BEFORE you start to draw your bar chart.

To do this you need to look at your data and find the biggest number of occurrences (i.e. the largest category).

With the traffic survey above this was not too difficult: the most cars in one category was 'cars with 1 person in' which had 4 cars.

This means that our highest number on the vertical axis is 4. We are dealing with discrete data and so the numbers on this axis will be 0, 1, 2, 3, 4 and the label will be 'Number of cars'.

> **Remember**
> Discrete data is data made up of things that are separate and can be counted.

We now need to decide on how many bars we are going to draw.

Again, this is already decided for us because we were looking at five categories in our survey, i.e.

- cars with 1 person in
- cars with 2 people in
- cars with 3 people in
- cars with 4 people in
- cars with 5 or more people in

and so we have **five** bars in our bar chart.

Once you have drawn your axes and labels you can now draw the bars as follows:

- Use a ruler.
- The height of each bar is the number of tallies you have for that category.
- The width of the bars must be equal.
- When you have finished drawing your bar chart don't forget to give it a title.

Section 1: Using data and statistical measures ■ Topic 2   HANDLING DATA

> **Method summary**
>
> - Make up a title for your bar chart.
>
> - Find out what the highest number of items is. This will give you the biggest number on the vertical axis (the one on the left going up). This will be the size of the tallest bar.
>
> - Decide how many bars to draw: this is the number of categories you are dealing with. The bars should be equal in width.
>
> - Draw and label your axes.
>
> - Using a ruler to draw your bars.

Activity 11

> The following data shows the number of flights from a regional airport on one day of the week made by different airlines.
>
> | Airline | Number of flights |
> |---|---|
> | Reily Air | 3 |
> | Easyfly | 4 |
> | English Airways | 1 |
>
> Draw a bar chart to display this data.

Check with the following suggestions before continuing.

Suggestions for Activity 11

Number of flights from a regional airport on one day of the week by different airlines

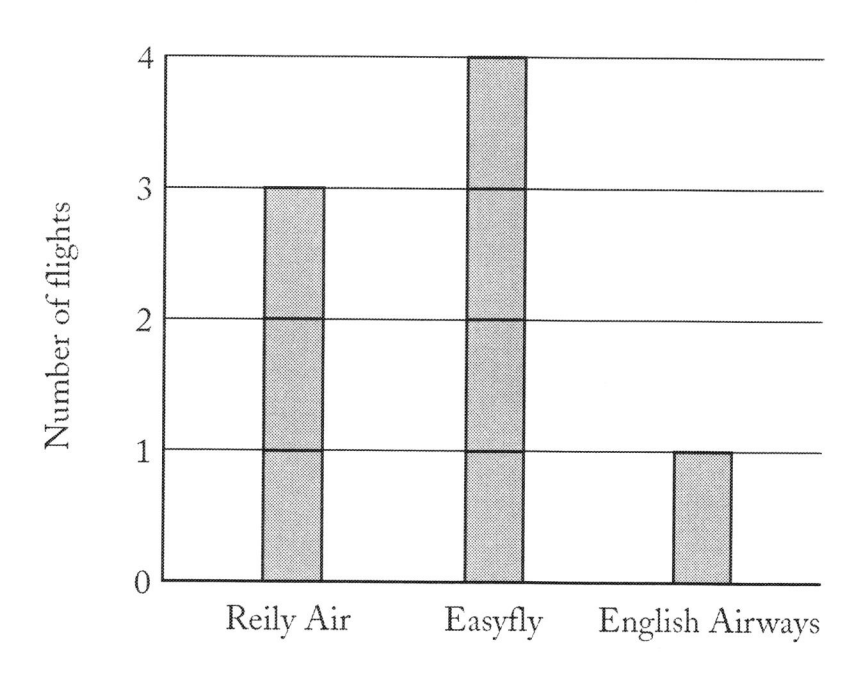

(Bar chart: Reily Air = 3, Easyfly = 4, English Airways = 1; y-axis: Number of flights)

Line graphs

Line graphs are drawn by marking (or plotting) points and then joining them with a straight line. You might have seen them used in holiday brochures or maybe on the television.

> **Hint**
> It is best to use graph paper when drawing graphs because it makes it easier to plot the points.

Section 1: Using data and statistical measures ■ Topic 2 HANDLING DATA

Example 11

The number of houses sold by an estate agent over a six-month period is shown below.

Month	Jan.	Feb.	March	April	May	June
No. of house sold	2	3	6	8	9	2

To draw this in a graph we need to first draw our axes. We will put the months on the bottom (horizontal) axis and the house sales going up the left side.

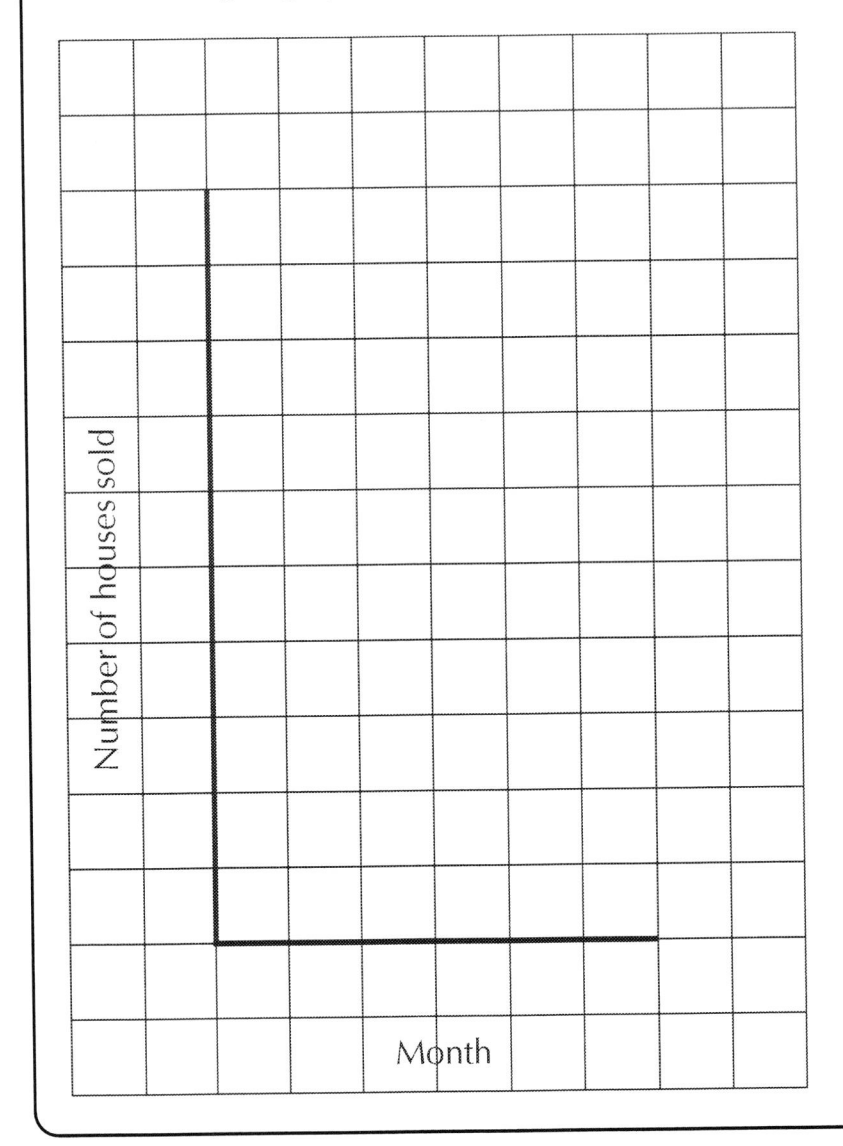

Number of houses sold

Month

Example 11 continued

Next we need to decide on how we divide up these axes.

- There are six months so there will be six marks on the bottom axis.
- The biggest number of house sales was 9 so the vertical axis will be from 0 to 10.

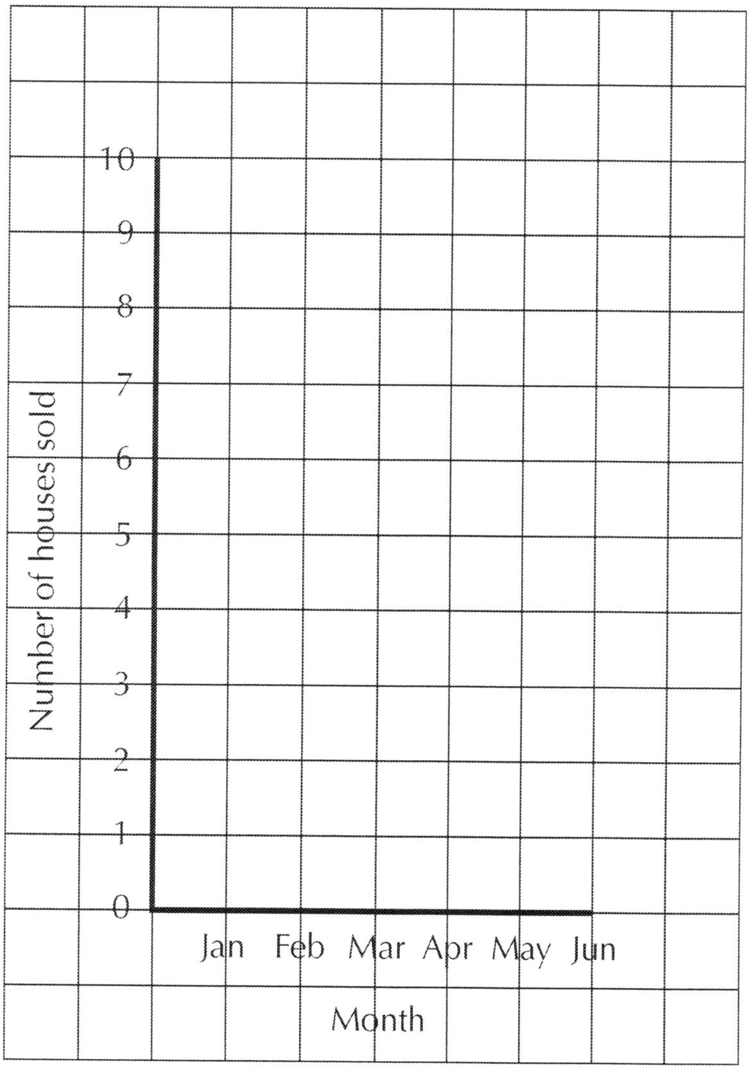
10
9
8
7
6
5
4
3
2
1
0
Number of houses sold
Jan Feb Mar Apr May Jun
Month

Section 1: Using data and statistical measures ■ Topic 2   HANDLING DATA

Example 11 continued

Now we can begin to mark our points. Starting with the number for January you go up the line marked 'Jan' and then when you get to the line going across marked '2' you make a small cross. You then do this for the other points.

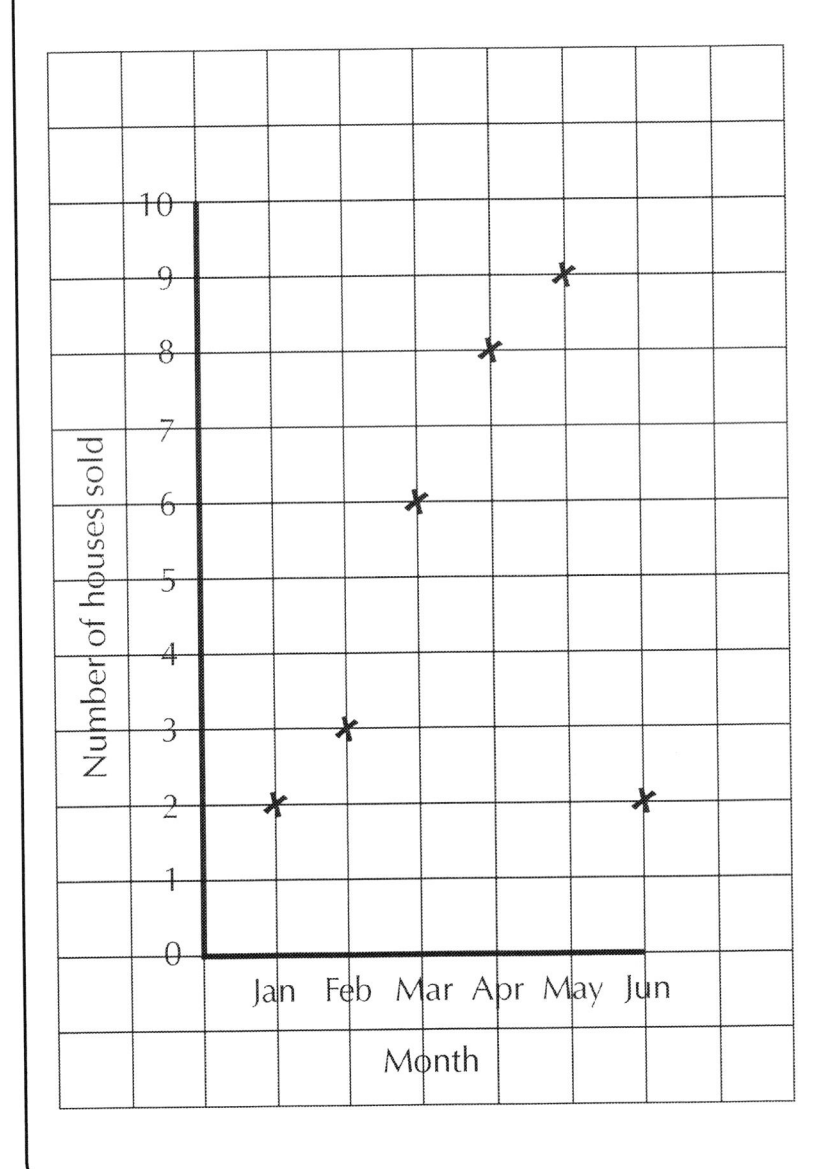

Example 11 continued

Finally, you join the points using a ruler and put in your title.

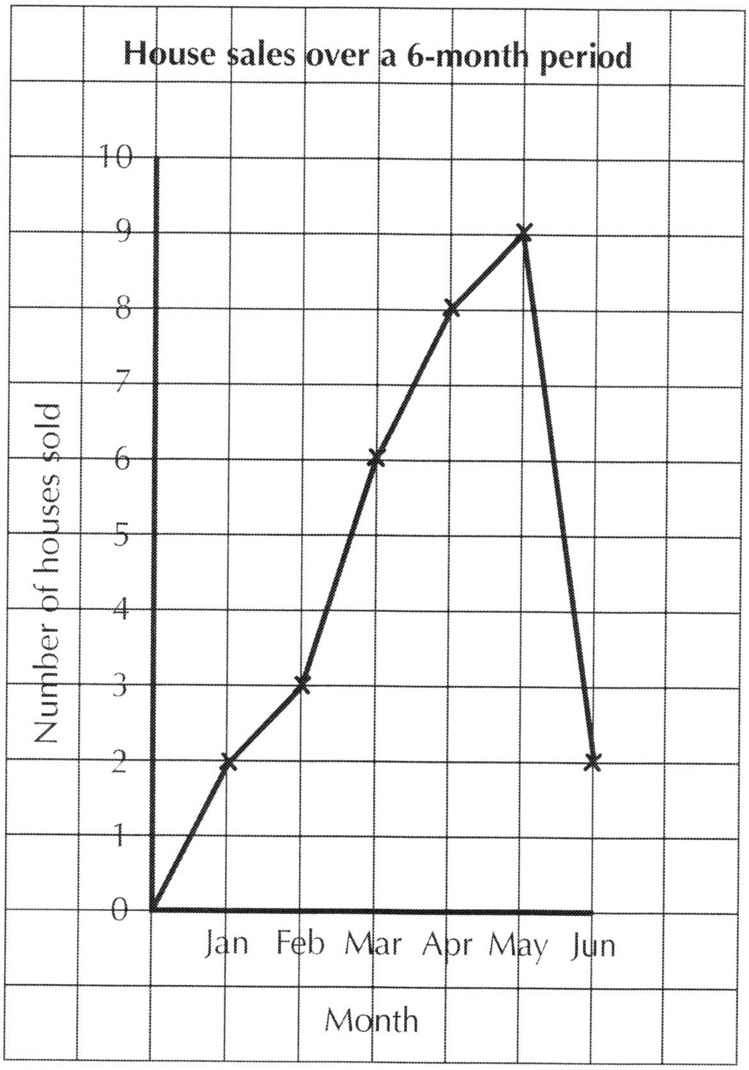
House sales over a 6-month period
Number of houses sold
10
9
8
7
6
5
4
3
2
1
0
Jan Feb Mar Apr May Jun
Month

Section 1: Using data and statistical measures ■ Topic 2 HANDLING DATA

> **Method summary**
>
> To draw a line graph you need to do the following:
>
> - draw the horizontal and vertical axes and label them
> - divide these axes into suitable scales. To do this you need to look at your data and find out what the smallest and largest numbers are
> - plot the points from your data (use a pencil and make small crosses)
> - join the points using a ruler
> - give your graph a title.

Activity 12

> In holiday brochures line graphs are often used to show resort temperatures or hours of sunshine.
>
> This is to tell you how hot or sunny a holiday resort should be. Here is a table showing the hours of sunshine at a holiday resort. Using the diagram below draw a line graph to display this data and then answer the questions below.
>
> | Month | May | June | July | August | Sept | Oct |
> |---|---|---|---|---|---|---|
> | Hours of sunshine | 6 | 7 | 8 | 9 | 8 | 7 |
>
> a) What month was the sunniest?
>
> b) What month had the least sunshine?

Check with the following before continuing.

HANDLING DATA — Section 1: Using data and statistical measures ■ Topic 2

Suggestions for Activity 12

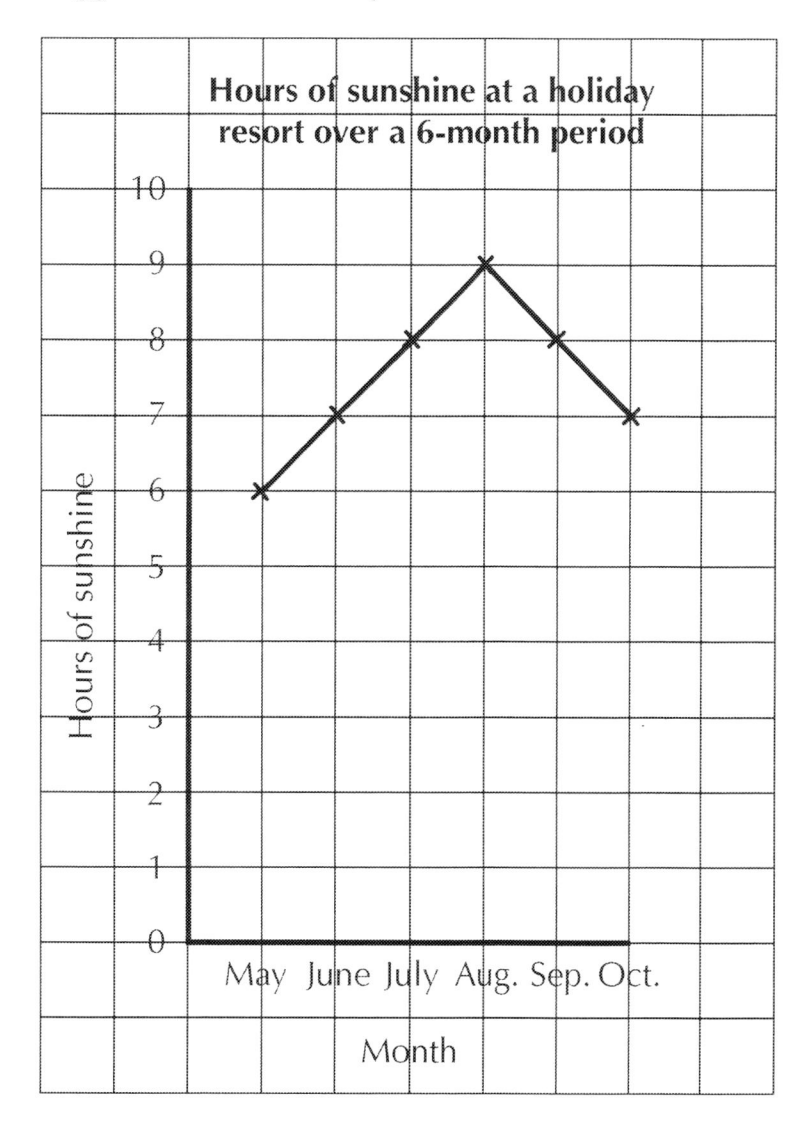

Hours of sunshine at a holiday resort over a 6-month period

(Graph showing hours of sunshine by month: May = 6, June = 7, July = 8, Aug = 9, Sep = 8, Oct = 7)

a) August

b) May

Self-check 8

Before moving on you need to make sure you are able to collect, organise and show data in the forms of tables, diagrams, charts and graphs. Ask yourself the following questions.

a) When I draw tables, diagrams, charts and graphs is my data displayed clearly so that the information is easy to understand?

b) Do I always put in titles, scales, labels and keys when they are needed?

If you are not sure about these points then show your work to someone else and ask if they understand the data.

Answers to Self-check 8 are at the end of the section.

Summary

In this topic you have:

- learned how to collect information in the form of tally charts

- displayed data using tables, diagrams, charts and graphs using sensible scales and clear labels

> By completing this topic you have covered HD1/L1.2 of the Data Handling section of the curriculum:
>
> - Collect, organise and represent discrete data (e.g. in tables, diagrams, charts and line graphs).

Topic 3: Finding the arithmetical average

So far in this section we have looked at handling data from the view of collecting, organising and showing information.

We are now going to look at processing data **numerically** rather than graphically.

This means that instead of displaying data in tables or pictures we will be using numbers.

We will doing some sums and finding one number to represent our data instead of lots of numbers.

In this topic we will do this by finding the arithmetical average. This may sound confusing but note that:

- arithmetical means 'doing sums'
- average means finding a middle value.

> **Note**
> With data we talk about 'data sets' or sets of data. Sets is just another word for 'group'. So, if we carried out a survey we would have a data set.

So when we say an arithmetical average we are talking about working out a middle value for our data with mathematical calculations.

Many will be familiar with the word 'average'. In fact, we may think it may mean 'not special' or 'just OK'. But that is not true in maths.

In maths it is just a middle value that should represent our data without us having to give all the details of the data.

In summary: by working out an average we can have one value that is representative of all our data and that uses all our data.

Where do we find averages in real life?

- If you look at a holiday brochure you will see that it will talk about 'average' hours of sunshine.
- A teacher might work out the average marks for students in a class.
- When we go on a journey we might talk about our average speed.

- With football you might work out the average goals scored per game over a season for your football team.

Finding the average

The arithmetic average is not difficult to work out. You need to do the following.

- Add up all your data (write down this total, let's say it makes 'A').

- Add up the number of 'bits' of data you have (let's call this amount 'B').

- Divide the total of your data by the number of bits of data (i.e. A ÷ B = average).

HANDLING DATA   Section 1: Using data and statistical measures ■ Topic 3

Example 12

The hours of sunshine per day during a week's holiday in June to Torquay was recorded as follows:

Sunday	6
Monday	1
Tuesday	7
Wednesday	8
Thursday	5
Friday	2
Saturday	6

We could draw a chart, a diagram or a line graph to present this data.

We can see, though, that as to be expected from the British weather, the amount of sunshine varied a lot.

We might, therefore, be more interested in the **average** amount of sunshine per day.

This would give us one value and an idea of how much sunshine to expect per day.

To work out this average value we would

- add up the amount of sunshine for each day and
- then divide this by the number of days we had the data for.

With this example we have:

$6 + 1 + 7 + 8 + 5 + 2 + 6 = 35$ hours of sunshine for the week

and 7 days of data.

So, the average is given by

$35 \div 7 = 5$ hours.

Note
You must remember what units you are working in and write in these units after your average value or it won't make sense.

So, from this data we can see that, on average, there were 5 hours of sunshine per day in a week in June in Torquay.

Section 1: Using data and statistical measures ■ Topic 3   HANDLING DATA

> **Note**
> Another word for 'average' is 'mean'. This must not be confused with, for example, 'mean' as being someone who doesn't want to spend money, or being unkind. Here 'mean' is a mathematical word meaning a type of average or middle value.

We could then use that information to help us choose our holiday.

For example, if we wanted more than 5 hours of sunshine a day for a holiday in June we would choose somewhere hotter (like Spain perhaps).

Method summary

- Add up all the bits of your data.
- Count the number of bits of your data.
- Divide the total of your data by the number of bits of data to give the average.
- Don't forget to put what units you are working in (e.g. hours, goals, people etc.).

Q. What are the advantages and disadvantages of using the arithmetic average?

A. Ever heard of families with 2.4 children? This is the national average but means nothing (because you can't have 0.4 of a child).

This highlights one of the problems with averages: the value you get may not be a real value in terms of what you are talking about.

Another problem is that the average value will be affected by the highest and lowest values.

For example, your football team could be having a really bad season and over nine games had scored nothing (so average goals scored would be 0), but they suddenly pick up and in the next match score 10 goals.

This would increase the average goals scored to 1 per match, which would make it looked like they'd scored a goal every match when they hadn't.

Averages are good, however, because they aren't too complicated to work out (compared to some other statistical calculations) and use ALL the data.

HANDLING DATA　　　　　　　　　　Section 1: Using data and statistical measures ■ Topic 3

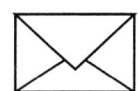

Investigation

Carry out an investigation with one of the following tasks.

EITHER

Work out the average age for people in your family or a group you belong to. Include as many members as you can. (To keep life simple with babies or small children just round up their ages to the nearest year.)

OR

Work out the average goals or points scored by your favourite sports team this season.

Now it's your turn to work out some averages in the following activities.

Activity 13

The ages of four children in a family are 4, 6, 8 and 10 years. Work out the average age.

Activity 14

Find the mean of the following data sets.

a) 4, 6, 11

b) 3, 7, 8, 4, 8

c) 8, 9, 10, 9, 4, 2

d) 11, 12, 13, 14, 15, 16

Activity 15

The number of goals scored by Fabfootball Rovers in recent matches were as follows:

2　3　0　1　3　2　3　2　1　3

Work out the average number of goals per match.

Check with the following suggestions before continuing.

Section 1: Using data and statistical measures ■ Topic 3 HANDLING DATA

Suggestions for Activity 13

7 years of age

Suggestions for Activity 14

a) 7

b) 6

c) 7

d) 13.5

Suggestions for Activity 15

2 goals per match

Self-check 9

a) In a maths class the scores for a test (out of 10) were as follows:

5	6	6	4	4
7	3	5	6	7
8	6	2	8	5
4	5	6	5	6

Work out the average score.

b) Some of the students felt the teacher had been too harsh with their marks. The tests were remarked and the new results were as follows:

4	6	6	4	4
6	1	5	6	6
7	6	1	9	5
3	5	6	5	5

Work out the average score for these new results and then answer the following question.

Which set of results gave the best marks? Was the teacher harsh with the first marking?

Answers to self-checks are at the end of the section.

Summary

In this topic you have:

- learned that the mean is one sort of average

- learned that the mean is worked out by adding up the items and dividing by the number of items

- understood that the mean can give a 'distorted average' if one or two values are much higher or lower than the other values.

> By completing this topic you have covered HD1/L1.3 of the Data Handling section of the curriculum:
>
> - Find the arithmetical average (mean) for a set of data.

HANDLING DATA   Section 1: Using data and statistical measures ■ Topic 4

Topic 4: Finding the range

Curr. ref. HD1/L1.4

In this topic we are going to continue looking at studying and processing data.

This will involve us finding the **range** for a set of data.

The range is the **difference** between the highest and lowest values in a set of data.

We work out differences (mathematically speaking) by subtracting (taking away).

Working out the range of a set can be good because it gives an idea of how large and varied a data set can be.

We talk about 'range' in real life in the following situations.

- Schools will have a range of ages of children.
- Companies will have employees on a range of salaries.
- Supermarkets have goods at a range of prices.

Finding the range

The first thing to do when finding ranges is to find the lowest and highest values in your data set.

You can do this in the following way.

- If your data set is not too big then the best thing to do is put the values in numerical order (lowest first).
- As you go through the data set tick or cross off the numbers as you put them in order so that you don't count the same one twice or miss one out altogether.

4̷ 2̷ 9 7 6 3̷ 5 8

2 3 4 .. ..

Once you have the highest and lowest values you then have to take the lowest away from the highest.

© National Extension College Trust Ltd   DEVELOP YOUR MATHS   67

This will give you the **range**.

The range measures the spread of a set of data. It is important because it can tell you how diverse your data is (or isn't).

Take, for example, the ages in a gardening club.

Say the average age is 40 years old. This doesn't tell you much about the people in the club.

- If the spread of the ages was 10 years then you know that everyone is either in their thirties or forties.

- BUT if the spread was 70 years then both youths and pensioners belong to the club.

So, the range gives you more information about a data set.

Remember, when working out the range you still have to include the units you are working in. So, if you are dealing with ages you will usually be talking about years and your range will be in years.

Example 13

Barry has 4 children. Sophie is 7 years, Karen is 4 years, Max is 12 years and Jason is 10 years. What is the range?

So here is our data set:

7   4   12   10

Let's put these numbers in order first:

4   7   10   12

It is easy to see that the lowest number is 4 and the highest is 12.

The range is worked out by taking the lowest away from the highest value, so:

Range = 12 − 4 = 8 years (don't forget to put the units [years] in).

In other words, the difference between Barry's oldest and youngest children is 8 years.

> **Method summary**
> - Write the numbers in numerical order (lowest first).
> - Find the lowest and highest numbers.
> - Take the lowest number away from the highest number to find the range for your data.
> - Don't forget to put what units you are working in (e.g. hours, goals, people etc.).

Let's try another example with a bigger data set.

Section 1: Using data and statistical measures ■ Topic 4   HANDLING DATA

Example 14

In Topic 3 you looked at the marks for a maths test. There were 20 scores in that data set. Here it is again:

5	6	6	4	4
7	3	5	6	7
8	6	2	8	5
4	5	6	5	6

To find the range we need to find the lowest and highest values.

We might be able to look at the data and see that the lowest number is 2 and the highest number is 8 but it is easy to get this wrong.

The best thing is to put the numbers in order (lowest first), ticking off the numbers as you go. Putting the numbers in order gives:

2	3	4	4	4
5	5	5	5	5
6	6	6	6	6
6	7	7	8	8

So, the lowest number is 2 and the highest number is 8.

To find the range we take the lowest from the highest and so

Range = 8 − 2 = 6 marks.

Now try the following activities.

Activity 16

Find the ranges for the following data sets

a) 1, 6, 7, 10

b) 7, 6, 2, 8, 10, 3, 11

c) 5, 4, 2, 8, 9, 11, 4, 12, 7

d) 5, 15, 6, 9, 12, 4, 2, 8, 1, 14

HANDLING DATA Section 1: Using data and statistical measures ■ Topic 4

Activity 17

In a random survey in Newcastle the ages of 20 people were:

61	18	42	37	32
15	25	52	74	23
49	41	58	31	42
21	27	65	47	35

a) Write the data set in order with the lowest number first.

b) What is the lowest age?

c) What is the highest age?

d) What is the range?

Check with the following suggestions before continuing.

Suggestions for Activity 16

a) Range = 10 − 1 = 9

b) Range = 11 − 2 = 9

c) Range = 12 − 2 = 10

d) Range = 15 − 1 = 14

Suggestions for Activity 17

a) 15 18 21 23 25
   27 31 32 35 37
   41 42 42 47 49
   52 58 61 65 74

b) 15 years

c) 74 years

d) Range = 74 − 15 = 59 years

Section 1: Using data and statistical measures ■ Topic 4   HANDLING DATA

Self-check 9

The lengths of 8 petals on a flower are measured as 56 mm, 45 mm, 63 mm, 66 mm, 47 mm, 53 mm, 59 mm and 64 mm.

Write the data set in numerical order and find the range of the data.

Self-check 10

The weekly earnings of 10 people employed in a call centre are as follows. Find the range.

| £210 | £195 | £187 | £196 | £201 |
| £205 | £198 | £188 | £183 | £204 |

Answers to the self-checks are at the end of the section.

Summary

In this topic you have:

- learned that the range measures the spread of a set of data.

- understood that the range is the difference between the smallest and largest values in a set of data.

By completing this topic you have covered HD1/L1.4 of the Data Handling section of the curriculum:

- Find the range for a set of data.

Suggested answers for self-checks

Self-check 1

a) Something like 'Drinks served during shift'

b) Tea; Coffee; Orange juice; Hot chocolate; Coke

c) Drinks; Number served

d) 5

e) 2 orange juices and 5 cokes = 7 cold drinks

f) 6 + 7 + 2 + 3 + 5 = 23

Self-check 2

a) 5 cm = 50 m

b) 4 cm = 40 m

c) 6 cm = 60 m

Self-check 3

Monday	Tuesday	Wednesday	Thursday	Friday
10	5	4	8	7

a) Monday

b) Wednesday

Self-check 4

a) Tesco

b) 7

c) Somerfield

d) ASDA = 6, Morrisons = 4. Together these make 10.

Self-check 5

a) A key and a title.

Section 1: Using data and statistical measures

b) Can be difficult to draw and can only show a few things.

c) Tells you how much a scale diagram or map has been 'shrunk'.

d) Titles tell you what tables, diagrams, charts and graphs are about and where and when the data was collected.

Self-check 6

No. of people in house	No. of responses
0	||||
1	⊬⊬ |||
2	⊬⊬ |
3	||

Self-check 7

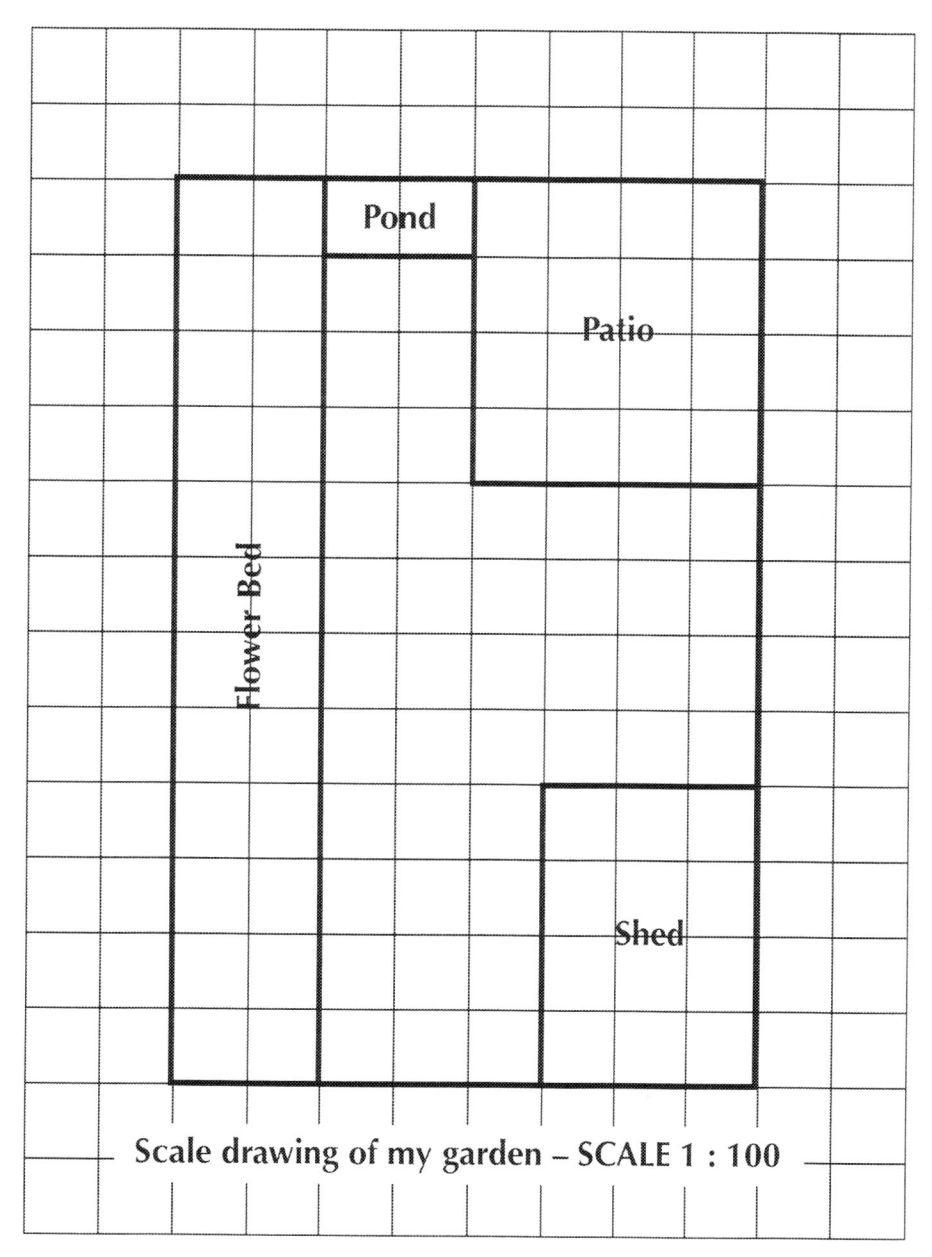

Scale drawing of my garden – SCALE 1 : 100
(showing Pond, Patio, Flower Bed, Shed)

Self-check 8

Before moving on make sure your tables, diagrams, charts and graphs are well drawn and laid out and easy to understand. Don't forget to put in titles, scales, labels and keys as needed.

Self-check 9

a) 5.4

b) 5

The first set in (a). The teacher had not been harsh.

Self-check 10

45 mm, 47 mm, 53 mm, 56 mm,

59 mm, 63 mm, 64 mm, 66 mm

Range = 66 − 45 = 21 mm

Self-check 11

Lowest is £183, the highest is £210.

Range is £210 − £183 = £27

Section 2: Probability

HANDLING DATA

Section 2: Probability

Probability is measuring how likely it is that something will happen.

We use probability in different ways in real life.

- Bookmakers use a form of probability to give betting odds (on anything).

- Insurance actuaries use probability to decide how much to charge for all the different types of insurance and assurance there is.

- Government departments use probability and statistics to help them govern the country.

Reasons for completing this section

Working through this section will enable you to:

- understand the possibility of different events happening

- show that some events are more likely to occur than others

- understand and use probability scales

- show the probability of events happening using fractions, decimals and percentages.

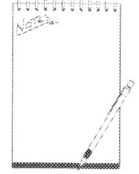

Resources

You will need:

- A pen
- Some paper
- A ruler
- A die
- A set of dominoes or a coin.

HANDLING DATA   Section 2: Probability ■ Topic 1

Topic 1: Introduction to probability

Curr. ref. HD2/L1.1

Probability is measuring how likely it is that something will happen.

Look at the word itself: 'probability'. Can you see it is related to the word 'probable'?

Another word for probability is **chance**.

You might say 'what are the chances of this happening?'.

For example, you might say 'I might cut the grass tomorrow'.

Probability would be used to measure how likely it is that you will cut the grass. There are two options involved here, either:

- you cut the grass
- or you don't.

If you knew that it was going to rain and you had lots of other things to do (and you hate cutting grass) then the **probability** of actually cutting the grass would be **low** or even **zero**!

But, on the other hand, if you really intended to cut the grass and the weather forecast was good then the **probability** of cutting the grass would be **high**.

We know that life is full of choices and chances.

We know that some things are more likely to happen than others.

We use probability to give us an idea of how likely it is that something will happen. It gives us a measuring system.

Let's investigate probability

We talked about cutting the grass earlier. We said the probability of cutting it was **high** or **low**.

This is one way of measuring (or rating) probability.

- If something is very likely to happen the probability is **high**.
- If something is not very likely to happen the probability is **low**.

Example 1

Say whether you think the probability of the following events happening is high or low:

a) Winning the National Lottery

b) Getting wet in the rain

c) Spring following Summer

In my opinion, the probability for these situations is as follows:

a) Low

b) High

c) High

Of course, some things have even chances of happening.

For example, if you toss a coin there is an equal probability of it being heads or tails.

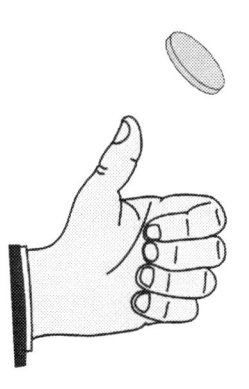

HANDLING DATA   Section 2: Probability ■ Topic 1

We would then say one of the following things:

- there is an even chance of it being heads or tails

OR

- there is a fifty-fifty chance of it being heads or tails.

How many different things that have different chances of happening can you think of?

Now try this activity.

Activity 1

Fill in the following table.

Events with high probability of happening	Events with even chance of happening	Events with low probability of happening

Check your answers with our suggestions before moving on.

Suggestions for Activity 1

As an example:

Events with high probability of happening	Events with even chance of happening	Events with low probability of happening
Moon rising tonight	Tossing a coin and getting heads	Winning the lottery
Death	Baby being a boy	Being kidnapped by aliens

Probability scales

In real life things usually fall somewhere in between the two extremes of

- definitely going to happen

and

- will never happen.

We can use a probability scale to measure how likely events are to occur.

On a probability scale

- certain events have a probability of 1
- all other events fall in between 0 and 1
- events with an even chance have a probability of $\frac{1}{2}$.

> **Note**
> You cannot have a probability bigger than 1 (certain) or smaller than 0 (impossible).

A probability scale looks like this:

```
impossible                                          certain
|───────────────────────┼───────────────────────|
0                      1/2                       1
```

To use a probability scale you mark on the line with a cross where you think the probability of something happening is.

So, for example.

- It is certain that June will follow July: on the probability scale you would put a cross on the line above the '1'.

- It is impossible, though, that cows will jump over the moon! So on the probability scale you would put a cross on the line above the '0'.

- The probability of a baby being a girl is about halfway: on the probability scale you would put a cross on the line above the '$\frac{1}{2}$'.

- For other events it is up to you to gauge the probability.

HANDLING DATA　　　　　　　　　　　　　　　　　Section 2: Probability ■ Topic 1

> **Memory box**
> - The probability of an impossible event is 0.
> - The probability of a certain event is 1.
> - All other events come between 0 and 1.

Now try this

Activity 2

> Use a ruler to draw your own probability scale. Mark on it 0, $\frac{1}{2}$ and 1. Label 'impossible' and 'certain'.
>
>
>
> Then mark these statements on the probability scale with crosses and label them with their question letter.
>
> a) The probability that the sun will rise tomorrow.
>
> b) The probability that I will run the London Marathon next year.
>
> c) The probability of dying one day.

Check your answers with our suggestions.

Suggestions for Activity 2

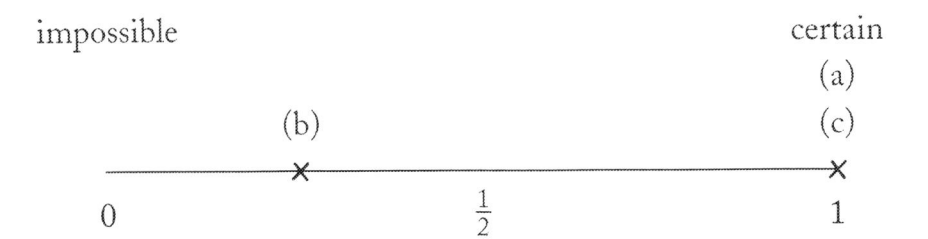

impossible   certain
                                                                  (a)
                        (b)   (c)
———————————×——————————————————————×
0                       $\frac{1}{2}$ 1

Of course, if you are a long distance runner (or plan to be one) your location for (b) might be closer to '1'!

Summary

In this topic you have:

- learned about the possibility of different events happening

- shown that some events are more likely to occur than others.

> By completing this topic you have covered HD2/L1.1 of the Data Handling section of the Adult core curriculum:
>
> - Show that some events are more likely to occur than others.

HANDLING DATA   Section 2: Probability ■ Topic 2

Topic 2: Measuring probability

In the previous topic we defined probability as 'the likelihood of an event happening'.

On a probability scale we gauged events as being

- impossible with a probability of 0
- certain with a probability of 1
- even with a probability of $\frac{1}{2}$.

In this topic you will further learn how we can measure probability using numbers.

Using fractions to write probabilities

In Topic 1 of this section we talked about tossing a coin.

In sport they often toss a coin to decide which team goes first or which direction they play.

We said the chance of getting heads or tails was equal or halfway on a probability scale.

This is the same as a probability of $\frac{1}{2}$.

We can work out probabilities by counting the number of possible outcomes and the number of successful outcomes.

> **Note**
> 'Outcome' is another word for 'result'.

Here we have only one head on a coin and there are two sides to a coin so there are two possible outcomes.

We could write this as:

$$p(\text{heads}) = \frac{1}{2} = \frac{\text{Number of heads}}{\text{Total number of possible outcomes}}$$

or, in the case of tails:

$$p(\text{tails}) = \frac{1}{2} = \frac{\text{Number of tails}}{\text{Total number of possible outcomes}}$$

Have you ever thrown a die hoping for a six?

Most of us have.

Provided the die wasn't loaded (i.e. designed to land on certain sides

Section 2: Probability ■ Topic 2

and give specific numbers), we know the probability of getting a six would be 1 in 6. Why is this?

There are **six** different numbers on a die:

1   2   3   4   5   6

The die can only land on **one** number.

So we have **1** successful outcome with **6** possible outcomes.

This is described as a 1 in 6 chance.

We could write this as:

p(getting a 6) = $\frac{\text{Number of successful outcomes}}{\text{Total number of possible outcomes}} = \frac{1}{6}$

What if you were allowed two throws of the die?

There are still 6 possible results, i.e. the numbers on the die

1   2   3   4   5   6

but this time there are 2 chances so there are 2 successful outcomes.

So we can write this as:

p(getting a 6) = $\frac{\text{Number of successful outcomes}}{\text{Total number of possible outcomes}} = \frac{2}{6}$

We can simplify $\frac{2}{6}$ to $\frac{1}{3}$ (because 2 and 6 are divisible by 2).

So with two throws of the die

p(getting a 6) = $\frac{1}{3}$

This makes sense because if you have two throws of the die then you are twice as likely to get a 6 and $\frac{1}{3}$ is double $\frac{1}{6}$.

HANDLING DATA   Section 2: Probability ■ Topic 2

Self-check 1

Before moving on make sure you understand the difference between possible outcomes and successful outcomes. Test yourself by filling in the following table.

Event	No. of successful outcomes	No. of possible outcomes
Throwing a six with a die		
Picking a red card from a pack of cards		
A pregnant woman expecting one child having a baby boy		

Answers to Self-check 1 are at the end of the section.

Now, try the next activity.

Activity 3

What if you had a special die which only had even numbers on it? That is, the numbers on the die were

2   4   6   2   4   6

You still want to throw a six and you have one throw to get it.

What is the probability of throwing a six?

Hint
The number of different numbers on the die is the number of possible outcomes.

Check your answer with our suggestion below.

Suggestion for Activity 3

There are 6 possible outcomes and 2 successful outcomes so the probability is $\frac{2}{6} = \frac{1}{3}$.

© National Extension College Trust Ltd   DEVELOP YOUR MATHS

Section 2: Probability ■ Topic 2   HANDLING DATA

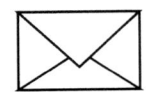

Investigation

Investigate one of the following.

Dominoes task

Using a set of dominoes find the following.

What is the probability of:

a) getting a domino with a double

b) getting a domino with a blank side

c) getting a domino which sides added together make more than 7?

OR

Coin tossing task

Copy and complete the following table for 20 tosses of a coin

	Number
Heads	
Tails	

Using your table what is the probability of getting

i) heads?

ii) tails?

How does this compare to your expected probabilities?

Method summary for finding probabilities

- Count the number of possible outcomes.

- Count the number of successful outcomes.

- Divide the number of events by the number of possible outcomes: $\dfrac{\text{Number of successful outcomes}}{\text{Total number of possible outcomes}}$

- Simplify this fraction if needed.

---

88   DEVELOP YOUR MATHS

HANDLING DATA

Section 2: Probability ■ Topic 2

Using decimals to write probabilities

We can write probabilities as decimals.

In the case of tossing a coin the probability of getting a head is 0.5. This is just $\frac{1}{2}$ written as a decimal.

To find probabilities in decimal form you use the same method as we used when writing probabilities as fractions, but you then convert the fraction to a decimal.

> **Remember**
> To convert fractions into decimals you divide the top part of the fraction by the bottom part. So here
> $\frac{1}{2} = 1 \div 2 = 0.5$

Example 2

> A bag contains 10 coloured cubes. Three cubes are red and 7 are green. What is the probability of selecting a green cube (in decimal form)?
>
> There are 10 possible outcomes (total number of cubes).
>
> There are 7 successful outcomes (number of green cubes).
>
> So the probability is
>
> p(green cube) = $\frac{\text{Number of successful outcomes}}{\text{Total number of possible outcomes}} = \frac{7}{10}$
>
> then convert this to a decimal
>
> $7 \div 10 = 0.7$
>
> So the probability of getting a green cube is 0.7.

> **Remember**
> You cannot have a probability bigger than 1 (certain) or smaller than 0 (impossible).

Activity 4

> What is the probability of selecting a red cube in the above bag (10 cubes in total with 3 red and 7 green)? Write the probability in decimal form.

Check your answer with our suggestions before moving on.

Suggestions for Activity 4

There are 10 possible outcomes and 3 successful outcomes.

The probability is $\frac{3}{10} = 0.3$ (in decimal form).

Now try the following Self-check question.

Section 2: Probability ■ Topic 2

Self-check 2

Kelly puts a coin into her purse. It now holds one £1 coin, two 10p coins and one £2 coin.

What is the probability (in decimal form) that the coin she put into her purse was:

a) a 10p coin

b) a £1 coin

c) a £2 coin?

The answers to self-checks are at the end of the section.

Using percentages to write probabilities

In addition to writing probabilities in fractions and decimals we can use percentages.

When we say that there is a fifty-fifty chance of something happening we are expressing probability using percentages.

Again, you use the same method as for working out probabilities using fractions and decimals but then change the decimal into a percentage.

Note
When we express probability using percentages the probability scale is changed slightly. 0 still means impossible but certain is now 100%. The probability must not be less than 0% or more than 100%.

Remember
To convert decimals into percentages you multiply the decimal by 100.
This is the same as moving the decimal point two places to the RIGHT.
So 0.5 x 100 = 50%
Don't forget to write the percentage sign (%).

HANDLING DATA   Section 2: Probability ■ Topic 2

Example 3

A letter is selected from the word

I M P O S S I B L E

What is the probability (in percentages) that the letter is a 'B'?

There are 10 letters in 'IMPOSSIBLE' and one 'B'.

So we have

p(getting a B) = $\frac{\text{Number of successful outcomes}}{\text{Total number of possible outcomes}}$ = $\frac{1}{10}$

We now convert this to a percentage

$\frac{1}{10}$ = 1 ÷ 10 = 0.1

0.1 × 100 = 10%

So there is a 10% chance of getting a B.

Now try this.

Activity 5

What is the probability (in percentages) of selecting an 'S' from the word

I M P O S S I B L E?

Check your answer with our suggestions before moving on.

Suggestions for Activity 5

$\frac{2}{10}$ = $\frac{1}{5}$ = 20%

Now try another activity.

Activity 6

A machine produces 10 bolts and 4 are found to be faulty.

What is the probability that a bolt picked up at random will NOT be faulty?

Check your answer with our suggestions before moving on.

Suggestions for Activity 6

If there are 4 faulty bolts then there are 10 − 4 = 6 bolts that are NOT faulty.

$\frac{6}{10}$ = 60%

Now try the final Self-check exercise.

Self-check 3

In this topic you have looked at writing probabilities using fractions, decimals and percentages.

Work out the following probabilities as fractions and then convert them to decimals and percentages.

Event	Probability		
	Fraction	Decimal	%
Rolling a die and getting a 1 or a 2			
Rolling a die and getting an odd number			
Rolling a die and getting a 4			

Answers to the Self-checks are at the end of the section.

Summary

In this topic you have:

- learned that the likelihood of an event is measured on a scale from 0 (impossible) to 1 (certain)

- learned that probability is expressed as the number of ways an event can happen divided by the total number of possible outcomes

- learned that probability can be written as a fraction, decimal or percentage.

> By completing this topic you have covered HD2/L1.2 of the Data Handling section of the Adult core curriculum:
>
> - Express the likelihood of an event using fractions, decimals and percentages with the probability scale of 0 to 1.

Section 2: Probability ■ Topic 2   HANDLING DATA

Suggested answers for self-checks

Self-check 1

Event	No. of successful outcomes	No. of possible outcomes
Throwing a six with a die	1	6
Picking a red card from a pack of cards	26	52
A pregnant woman expecting one child having a baby boy	1	2

Self-check 2

a) $\frac{2}{4} = \frac{1}{2} = 0.5$

b) $\frac{1}{4} = 0.25$

c) $\frac{1}{4} = 0.25$

Self-check 3

Event	Probability Fraction	Decimal	%
Rolling a die and getting a 1 or a 2	$\frac{1}{3}$	0.33	33%
Rolling a die and getting an odd number	$\frac{1}{2}$	0.5	50%
Rolling a die and getting a 4	$\frac{1}{6}$	0.166	16.6%

Glossary

Glossary

The following terms all feature in the 'Handling data' workbook. As they are key terms, they appear in the text in **bold**. The definitions relate to the use of each term in this workbook.

angle	a form of measuring the space between two lines that touch; usually measured in degrees (°)
average	the **mean** across several records (e.g. the average amount of sunshine per day over a period of a month)
bar chart	a particular form of representation of data. Data are represented by bars of equal width where the lengths are proportional to the data value. The bars may be shown vertically or horizontally
chance	another word for **probability**
charts	'maths pictures'
data	information of a quantitative nature consisting of counts or measurements
data set	data that has been collected is called a 'data set'; a group of data items
decimal	relating to base 10. Most commonly used in relation to tenths, hundredths, thousandths being shown as digits following a decimal point
diagrams	pictures that use a figure to represent something
difference	the amount by which one figure is greater or lesser than another
discrete data	data resulting from a count of separate items or events (e.g. a number of people)
event	used in probability to describe the outcome of an action or a happening

fraction	when a whole number or object is divided into equal parts (e.g. 1 divided by 8 gives $\frac{1}{8}$)
graphs	a form of representation of grouped data
line graphs	diagrams showing a relationship between two variables
map	a diagram or plan showing the layout of an area, whether it is a room, country or road system
mean	a type of average. The arithmetic mean is the sum of quantities divided by the number of them
percentage	a fraction expressed as the number of parts per hundred and recorded using the symbol %
pictograms	a particular way of representing data using symbols to show objects. Such symbols can mean either one object or many objects as defined in the key
pie chart	a circle cut into pieces. The size of each piece of the 'pie' is proportional to the angle at the centre of the circle
probability	the likelihood (or chance) of an event happening. Probability is expressed on a scale from 0 to 1. When an event is impossible it has a probability of 0; when an event is certain it has a probability of 1. All other events fall somewhere in between 0 and 1
protractor	a device used to measure and draw angles
range	the difference between the biggest and smallest numbers in a data set
real life	the size of things as they actually are (as opposed to things drawn to scale)
scale	the amount a map or plan has been reduced
scale drawing	a plan or map in which distances and objects are reduced by the same amount, drawn to **scale**

tables	a method of organising and representing data using rows and columns
tally chart	a way of collecting data using a table and a series of tallies (where I = 1 etc.)
zero	the number 0 (meaning nothing)

Information on the National Test

Information on the National Test

How is adult numeracy assessed?

To gain the Certificate in Adult Numeracy at Level 1 you will need to pass a test. The test lasts for one hour and 15 minutes. It is a multiple-choice test covering handling data, number and measures, shape and space.

There are 40 questions and you are given a choice of four possible answers for each question. You have to pick out the right answer and mark it on an answer sheet (or click the answer if you take the test online). You are not allowed to use a calculator during the test.

Entering for the test

You can take the Certificate in Adult Numeracy at various times during the year at a registered test centre.

If you are working through this series on your own, you will need to contact a test centre to enter you for the test. Your local college should be able to help you. If not, you can phone **learndirect** on 0800 100 900. They will be able to tell you about centres near you.

Make sure you register for the test in plenty of time. Some centres will want you to give two months notice. You may be charged a fee to take the test. Be sure to ask your test centre about this.

Additional help to access assessment

If English is not your first language, or if you need help reading the test paper, you must tell the centre when you register for the test. You may be able to take a dictionary with you, or have someone to help you read the paper.

If you need the centre to make any other arrangements so that you can take the test, you must tell them when you register. They may need to make some checks or they may need time to make the arrangements.

Preparing for the test

- You need to complete all three workbooks in this series (*Handling Data*, *Number* and *Measures, Shape & Space*) or be confident in the skills covered before taking the test.

- Make sure you know the date and the time of your test.

- Look back through the workbooks and the activities you have done.

- Try some multiple-choice tests. Ask the centre if they can provide you with any practice test papers. Alternatively, try the Key Skills example tests on the QCA website (www.qca.org.uk).

Taking a test can be scary but remember you have worked hard to complete the workbooks. The test is giving you the opportunity to prove how much you know.